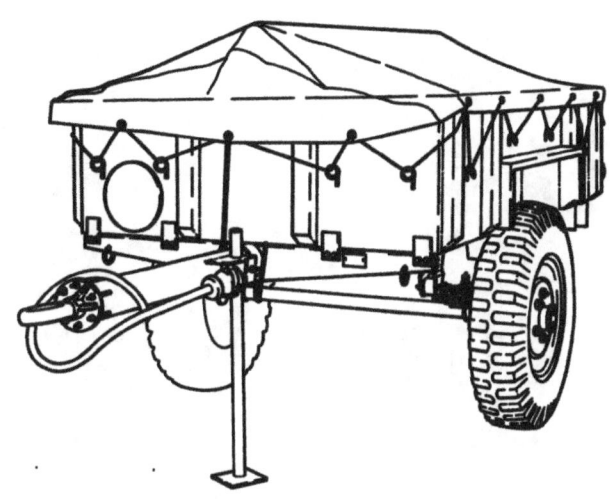

REPAIR PARTS SCALE

02060

2330-66-049-4263

TRAILER, CARGO ½ TON, 2 WHEELED, AUST NO 5

ISSUE 4   FEB 01

CONTENTS

SUBJECT	PAGE
TITLE PAGE	I
CONTENTS	II
PREFACE	III
REPAIR PARTS SCALES REQUEST	V
SCALES CHANGE REQUEST	VII

CHAPTER 1 – INDEXES

SECTION	TITLE
1	REPAIR PARTS SCALES NIIN INDEX
2	REPAIR PARTS SCALES GROUP INDEX
3	IDENTIFICATION PLATE

CHAPTER 2 – TEXT

1	REPAIR PARTS SCALES – TEXT AND ILLUSTRATIONS

PREFACE

Purpose of the Repair Parts Scale

1. This Repair Parts Scale (RPS) is an entitlement document prepared by Defence Materiel Organisation (DMO), Technical Data Centre (TDC). It authorises Units, Technical Support Sub-Units and Workshops that have the responsibility, the trade qualified personnel and the facilities, to draw Repair Parts for the repair and maintenance of the equipment detailed within the RPS.

Basis of Scale

2. The RPS is assessed on the premise that a repair part is any essential component or accessory which, through deterioration, breakage, normal wear and tear or loss, is likely to require replacement during the expected life of the equipment, and is an item which could not be economically repaired or fabricated in workshops.

3. The range and breadth of repair parts listed in the RPS is based on the grades of repair authorised for the equipment.

Recommended Stocking Quantities

4. The scale or depth of repair parts (also known as stocking quantities) are not listed in the RPS.

5. A Recommended Stocking Quantities report for units or force scaling for deployments and exercises may be obtained from The **RPS/CES Manager** using the same Web, E-mail, Fax and Mail addresses as listed under General Enquiries.

6. The following information is to be supplied with each Stocking Quantity Report request:

   a. RPS Number(s);

   b. equipment asset NATO Stock Number(s) (NSN) or Entitlement Group Code(s) (EGC);

   c. equipment asset name(s);

   d. number of equipment to be supported;

   e. duration of the support in months;

   f. the grade of repair at which the support is to be provided ie Light (L), Medium (M) and Heavy (H);

   g. type of forecast required if more than one equipment is involved (ie a single forecast per equipment or a consolidated report listing all equipment in one report); and

   h. the date the report is required by.

Access to RPS On-line

7. RPS can be accessed for viewing on-line at **http://vbmweb.sor.defence.gov.au/techdata** The RPS on this site can be Printed and Saved but when printed or saved they are Uncontrolled Copy.

Additional Copies of RPS or RPS Index

8. Personnel authorised by their unit to request additional copies of RPS or the RPS Index can do so using the same Web, E-mail, Fax and Mail addresses as listed under General Enquiries.

New Issues and Distribution

9. New RPS and new Issues of RPS are distributed automatically to authorised recipients.

10. When a new Issue of an RPS is released it supersedes all previous Issues, copies of which should be destroyed.

Delivery Address for RPS

11. Unless special circumstances exist, RPS or Indexes will not be addressed or sent to individuals they will be delivered to the Unit or Sub Unit address (eg. 7 Sig Regt or 3 RAR Tech Spt). Personnel requesting RPS or Indexes are to ensure they are authorised to submit the request and that the receiving Unit or Sub Unit will complete the internal delivery.

General Enquiries

12. For general enquiries in respect of any RPS contact The RPS/CES Manager via the TDC website address as per Para 7.

by e-mail to:
LEATechnicalData@drnex.defence.gov.au

by Telephone to: (03) 9282 7812

by Facsimile to: (03) 9282 7618 or

or, by Mail to:

RPS/CES Manager
Technical Publications Section
Technical Data Centre
DPM-3
661 Bourke Street
MELBOURNE VIC 3001

Changes to RPS and Scales Change Requests

13. Amendments to RPS will normally be made in the form of new issues.

14. Users may request changes to RPS by submitting Scales Change Requests (SCR) using the same Web, E-mail, Fax and Mail addresses as listed under General Enquiries.

15. Hardcopy of SCR proformas are in the front of every RPS. Electronic copies can be submitted through the TDC website address as per Para 7.

16. Scales Change Requests may be raised by:

a. units, repair elements, logistic units and headquarters who find, through user experience, the scale is inadequate either by range or quantity of repair parts;

b. units, repair elements, logistic units and headquarters who consider previously non-scaled items could become a recurring requirement; or

c. personnel who have identified errors or omissions in the text or illustrations.

17. Scales Change Requests will be acknowledged by TDC. If a change is approved, then users will be notified in the form of interim (pen) amendments. The changes will then be incorporated in the next issue of the RPS

18. It is emphasised that units, repair elements and logistic units are free to adjust their holdings as soon as user experience indicates the need. Adjustment action does not require prior approval of a Scales Change Request. The latter is required by TDC so the RPS database may be updated to reflect the latest information. This data may then be used for subsequent stocking quantity reports, reflecting the latest usage data.

How to Use the RPS

19. Refer to the Group Index in the Contents page of the RPS to locate the starting page number of the group required.

20. Refer to the group illustration and identify the item required by its callout number.

21. Refer to the text page, the callout number of the item required is listed in the column titled ITEM NO. Information for each item is listed in the columns titled:
DESIGNATION,
NSN
MANUFACTURER CODE / PART NO
SUPPLIER CODE / PART NO,
NO OFF (per assembly),
UOI (unit of issue),
EXP (stock type), and
L* M* H* (grade of repair details.)

22. The RPS Indexes also allow users to locate the position of parts for which the NSN or NATO Item Identification Number (NIIN) is known. Refer to the NIIN sequence index to locate the group and callout number of the item.

Illustrated Items

23. All illustrated items have an accompanying Item number in column 1 of that group's text page; eg AAB 007 indicates callout number 7 on the illustration for group AAB.

Non-illustrated Items

24. All non-illustrated items within a group are identified by having an Item number of 901 (formerly 9001) or greater; eg AAB 901 indicates the first unillustrated item within group AAB, AAB 902 the second and so on. The non-illustrated item numbering system has been changed to comply with the requirements of the Standard Defence Supply System (SDSS).

Scaled Items

25. Scaled items are indicated as such by having the letters "L", "M" or "H" in the grade of repair column, and a NSN in the NSN column.

26. In some instances, particularly for first issue RPS, items intended to be scaled may not have a NSN listed but will be identified for repair by "L", "M" or "H" in the grade of repair column. As soon as cataloguing action is completed for these items, a new issue of the RPS will be released.

Non-scaled Items

27. Non-scaled items have the grade of repair column blank, do not have an NSN and therefore do not appear in the NIIN index.

28. Requests for non-scaled items are to be carefully examined by units' technical support sub-units, workshops and contract repair elements. Requests for non-scaled items should be directed to the applicable National Fleet Manager.

Manufacturers' and Suppliers' Codes and Part Numbers

29. Prime manufacturers' codes and part numbers, if available, will appear directly below the NSN field. If not available, the field will read "NIL" or be blank. The next higher assembly should be requested for items with "NIL".

30. Suppliers' codes and part numbers, if available, will appear directly below the prime manufacturers' code and part number field. If not available, the field will be a duplicate of the manufacturer's code and part number.

INTENTIONALLY

LEFT

BLANK

REPAIR PARTS SCALES REQUEST

UNIT TITLE AND ADDRESS:	YOUR REFERENCE NO:
	UNIT CONTACT TELEPHONE NO: STD: DNATS:

PLEASE SUPPLY THE FOLLOWING REPAIR PARTS SCALES

RPS NO.	ISSUE NO.	REPLACEMENT COPIES QTY (1)	AUTOMATIC DISTRIBUTION QTY (2)	REMARKS

RAISED BY:

....

NAME (PRINT)   SIGNATURE   RANK   APPOINTMENT   DATE

Notes: 1. In this column, indicate the number of copies you require now (this does not affect automatic distribution).
2. In this column, indicate the number of copies required for automatic distribution of future issues of this RPS.

TECHNICAL DATA CENTRE, PUBLICATIONS SECTION USE ONLY

DATE RECEIVED: .... DATE ACTIONED: ....

PLEASE STAPLE OR TAPE BEFORE MAILING

OHMS

DEPARTMENT OF DEFENCE (ARMY)

If not delivered, return to:

**RPS/CES MANAGER
PUBLICATIONS SECTION
TECHNICAL DATA CENTRE
DPM-3
661 Bourke Street
MELBOURNE VIC 3001**

FOLD HERE

SCALES CHANGE REQUEST

UNIT TITLE AND ADDRESS:	YOUR REFERENCE NO:
	UNIT CONTACT TELEPHONE NO: STD: DNATS:

RPS NO:	ISSUE NO:	ISSUE DATE:

GROUP	ILLUS NO.	DETAILS

RAISED BY:

....

NAME (PRINT)   SIGNATURE   RANK   APPOINTMENT   DATE

TECHNICAL DATA CENTRE, PUBLICATIONS SECTION USE ONLY

ENTERED DATABASE [____]

UNIT SERIAL NO [_____]   DATE RECEIVED ____/____/____   USER ID [_____]

OP21-F10   Issue 1

PLEASE STAPLE OR TAPE BEFORE MAILING

OHMS

DEPARTMENT OF DEFENCE (ARMY)

If not delivered, return to:

RPS/CES MANAGER
PUBLICATIONS SECTION
TECHNICAL DATA CENTRE
DPM-3
661 Bourke Street
MELBOURNE VIC 3001

FOLD HERE

REPAIR PARTS SCALE DISTRIBUTION LIST

UNIT	QTY	UNIT	QTY
TECH DATA CENTRE (RPS MGR)	1	11 BASB FD WKSP COY	2
ESLMD	1	13 BASB WKSP COY	3
LEA (EIO)	1	10 TPT SQN (13 BASB)	1
DNSDC - TECH LIBRARY	5	NLG (BULIMBA)	3
DNSDC S/H 16	1	NLG - TOWNSVILLE (WORKSHOP)	1
DNSDC - PENRITH	1	SLG (PUCKAPUNYAL)	3
DNSDC - ACT WKSP PL	2	SLG (HOBART)	3
DNSDC HUNTER VALLEY DET	3	JLU - NTH LOG SPT COY	2
DNSDC - CATC WKSP	2	JLU - STH LOG SPT COY	2
10 FSB	2	JLU - WEST LOG SPT COY	2
10 FSB WKSP	4	1 ARMD REGT (EME OPS)	8
10 FSB VEH PLANNER	3	2 CAV REGT TECH SPT SQN	5
2 FSB (LSF WKSP)	3	3/4 CAV REGT TECH SPT	2
1 CSSB WKSP	3	2/14 LHR (QMI) TECH SPT SQN	5
1 CSSB SUP COY	2	3/9 SAMR A SQN TECH SPT	1
7 CSSB WKSP	5	4/19 PWLH TECH SPT	2
7 CSSB TPT SPT PL	2	10 LH A SQN TECH SPT TP	1
11 BASB	2	12/16 HRL TECH SPT	1
3 BASB FD WKSP	3	1 RAR TECH SPT	2
3 BASB FIELD SUPPLY COY	1	2 RAR TECH SPT	3
4 BASB	2	3 RAR TECH SPT	2
4 BASB WKSP COY	3	5/7 RAR TECH SPT	5
4 FD SUP COY (4 BASB)	1	6 RAR TECH SPT	2
5 BASB(WKSP,FD SUP,TPT COY)	3	9 RQR	2
8 BASB SUP COY	1	1/15 RNSWL TECH SPT	3
8 BASB WKSP COY	4	4/3 RNSWR TECH SPT	2
9 BASB WKSP COY	2	10/27 RSAR TECH SPT	1

ISSUE NO: 4   REPAIR PARTS SCALE: 02060   PAGE NO: 1

REPAIR PARTS SCALE DISTRIBUTION LIST

UNIT	QTY	UNIT	QTY
12/40 RTR	2	38 SPT SQN (4 CER)	2
41 RNSWR TECH SPT	1	1 JSU TECH SQN	2
25/49 RQR TECH SPT	2	145 SIG SQN	2
51 FNQR TECH SPT	1	DEFENCE COMMS ELEMENT-QLD	2
1 COY, 1 CDO REGT	2	DCE VIC (138 SIG SQN)	2
2 COY, 1 CDO REGT	1	7 SIG REGT (EW)	2
PILBARA REGT TECH SPT	1	8 SIG REGT WKSP	3
NORFORCE TECH SPT	1	1 CSU (COMMAND SPT UNIT)	2
MONASH UNIVERSITY REGIMENT	1	108 SIG SQN	1
1 FD REGT TECH SPT	3	126 SIG SQN	1
2/10 MDM REGT TECH SPT	2	DCS - CANBERRA	1
4 FD REGT TECH SPT	2	7 CSU CIS SQN	1
7 FD BTY TST	2	152 SIG SQN	1
7 FD REGT TECH SPT	3	15 TPT SQN TECH SPT TP	2
8/12 MDM REGT TECH SPT	2	26 TPT SQN WKSP	3
16 FD BTY TECH SPT	1	2 FSB (44 TPT TECH SPT)	1
23 FD REGT TECH SPT	1	ALTC - TRAINING FLEET	2
48 FD BTY TECH SPT	1	5 AVN REGT TECH SPT	4
CATC HQ	1	162 RECCE SQN SPT TP	3
CATC (OFFENSIVE SPT DIV)	2	LAND COMD BATTLE SCHOOL	2
2 CER TECH SPT	3	ACOMMS TC (FD TECH TROOP)	2
1 CER TECH SPT	2	176 AIR DISPATCH SQN	2
3 CER WKSP TP (HQ SQN)	2	AVN SPT GP WKSP	2
17 CONST SQN TECH SPT	3	NLG - WALLANGARRA	2
21 CONST SQN TECH SPT	2	MYAMBAT LOG COY	1
13 FD SQN (13 CER)	2	1 MP COY	2
35 FD SQN (11 CER)	1	2 MP COY (22 MP PL) (5 BASB)	2

REPAIR PARTS SCALE DISTRIBUTION LIST

UNIT	QTY	UNIT	QTY
3 MP COY	1	16 RWAR TECH SPT	1
ARMY MIL POLICE TRG CNT	1	1 CER	1
ALTC - BONEGILLA (HEALTH)	1	5 CER	1
ARMY RECRUIT TRAINING CENTRE	1	85 TPT TP (TRL)	1
REGIONAL TRG CNT (NSW)	2	1/19 RNSWR TECH SPT	3
PARACHUTE TRG SCHOOL	1	142 SIG SQN	3
P&EE GRAYTOWN	1	ALTC-VEHICLE WING	3
P&EE PT WAKEFIELD	1	RMC	1
HQ COY 1 BDE TECH SPT	1	4/19 PWLH (RECON)	1
DCSO(R)	1	SASR TECH SPT	3
HQ COY 3 BDE TECH SPT	2	141 SIG SQN	1
7 CSU ADMIN SQN	1	ANSCE TANAGER (ASJ42)	6
HQ 7 TF (EME)	1	A FD BTY TECH SPT	2
HQ 1 CDO REGT	1	TOTAL QUANTITY	301
AMTDU (ARMY COMPONENT)	1		
31 RQR	1		
4 MP COY	1		
1 HEALTH SUPPORT BN (WKSP)	3		
UNIVERSITY NSW REGT	2		
11 FD SQN (7 CER)	2		
14 CES (8 CER)	1		
1 TRG GP TECH SPT SECT	1		
1 AVN REGT (TASS)	6		
4 RAR TECH SPT	2		
DCFO-CANUNGRA WKSP	1		
161 RECCE SQN	3		
145 SIG SQN WKSP	3		

INTENTIONALLY LEFT BLANK

CHAPTER 1

SECTION 1

REPAIR PARTS SCALES
NIIN INDEX

INTENTIONALLY LEFT BLANK

REPAIR PARTS SCALE NIIN INDEX

NIIN 99-999-9999	GROUP IDENT NO		NIIN 99-999-9999	GROUP IDENT NO		NIIN 99-999-9999	GROUP IDENT NO	
00-100-3872	B	004	66-017-0071	B	008	66-097-0325	AD	004
00-679-4584	B	009	66-017-0072	B	013	66-097-0434	AA	902
00-959-1488	AB	006	66-017-0073	EA	901	66-097-0964	AB	901
01-411-3898	GA	001	66-017-0074	EA	013	66-098-4307	GA	009
12-128-1833	GA	002	66-017-5137	FA	003	66-098-4413	GA	007
12-188-1750	GA	013	66-017-5149	AD	001	66-107-8325	AB	001
66-010-7012	D	004	66-017-5150	AB	010	66-108-0508	D	001
66-010-7415	EA	005	66-017-5338	EA	011	66-116-5535	GA	902
66-010-7421	GA	004	66-017-5340	EA	009	66-120-5419	GA	011
66-010-7460	GB	006	66-017-9989	GA	905	66-122-8449	AA	903
66-010-7864	D	002	66-019-0931	AA	004	66-122-8827	B	003
66-010-9168	FA	007	66-019-3656	AA	010	66-122-9246	B	010
66-010-9530	EA	002	66-019-3680	FA	009	66-122-9502	GA	012
66-011-0004	FA	011	66-019-3719	B	015	66-132-7969	AB	004
66-011-0776	GB	001	66-019-3763	EA	907	66-143-7496	GB	030
66-011-9348	FA	902	66-019-3767	EA	001	66-143-7497	GB	024
66-012-3974	FA	015	66-019-3897	EA	010	66-143-7498	GB	032
66-012-5751	GB	013	66-019-3956	AA	005	66-143-7499	GB	027
66-012-8299	FA	901	66-019-3956	EA	909	66-147-0583	GA	906
66-013-0067	AA	009	66-019-3957	EA	903	66-147-0619	AA	901
66-013-0067	EA	908	66-019-3959	EA	003	99-128-7539	GB	012
66-013-0067	GA	003	66-019-3960	EA	007	99-406-7607	D	003
66-013-0070	B	014	66-021-0191	EA	012	99-803-3383	GB	008
66-013-0070	EA	902	66-021-0731	GB	018	99-803-3384	GB	009
66-013-3391	AA	003	66-024-5131	FA	012	99-803-3390	GB	005
66-013-9530	FA	008	66-025-2431	GA	006			
66-014-8509	GB	019	66-027-7458	GB	020			
66-015-5038	GB	901	66-036-6131	EA	905			
66-016-0720	GA	016	66-036-6132	EA	904			
66-016-0815	FA	006	66-036-6133	EA	906			
66-016-5980	EA	006	66-036-9096	B	012			
66-016-5993	EA	008	66-036-9099	B	901			
66-016-5995	FA	010	66-038-0295	AB	008			
66-016-5997	B	902	66-038-0295	B	016			
66-016-8516	GA	015	66-043-0011	GB	028			
66-016-8932	AB	003	66-044-0246	AB	902			
66-016-9984	AB	005	66-044-6203	GA	904			
66-016-9985	AD	005	66-044-6204	GA	903			
66-016-9986	AD	002	66-049-4073	AB	002			
66-016-9987	B	002	66-060-0704	FA	014			
66-016-9988	B	007	66-068-5743	FA	013			
66-016-9990	B	011	66-090-7823	GB	016			
66-017-0060	AB	007	66-091-9350	GA	005			
66-017-0061	B	001	66-092-6259	GB	022			
66-017-0070	FA	002	66-093-2784	GA	014			

ISSUE NO: 4   REPAIR PARTS SCALE: 02060

INTENTIONALLY LEFT BLANK

CHAPTER 1

SECTION 2

REPAIR PARTS SCALES
GROUP INDEX

INTENTIONALLY LEFT BLANK

REPAIR PARTS SCALE GROUP INDEX

GROUP	CONTENTS	START PAGE
AA	FRAME ASSEMBLY	1
AB	TOWING ATTACHMANT FRONT AND REAR	3
AD	STAND LEVELLING	5
B	AXLE ASSEMBLY	6
D	WHEEL, RIM AND TYRE	8
EA	SPRINGS AND MOUNTINGS	9
FA	CARGO BODY AND MOUNTINGS	11
GA	MAIN WIRING HARNESS	13
GB	LIGHTS	16

INTENTIONALLY

LEFT

BLANK

CHAPTER 1

SECTION 3

IDENTIFICATION PLATE

INTENTIONALLY LEFT BLANK

IDENTIFICATION 02060

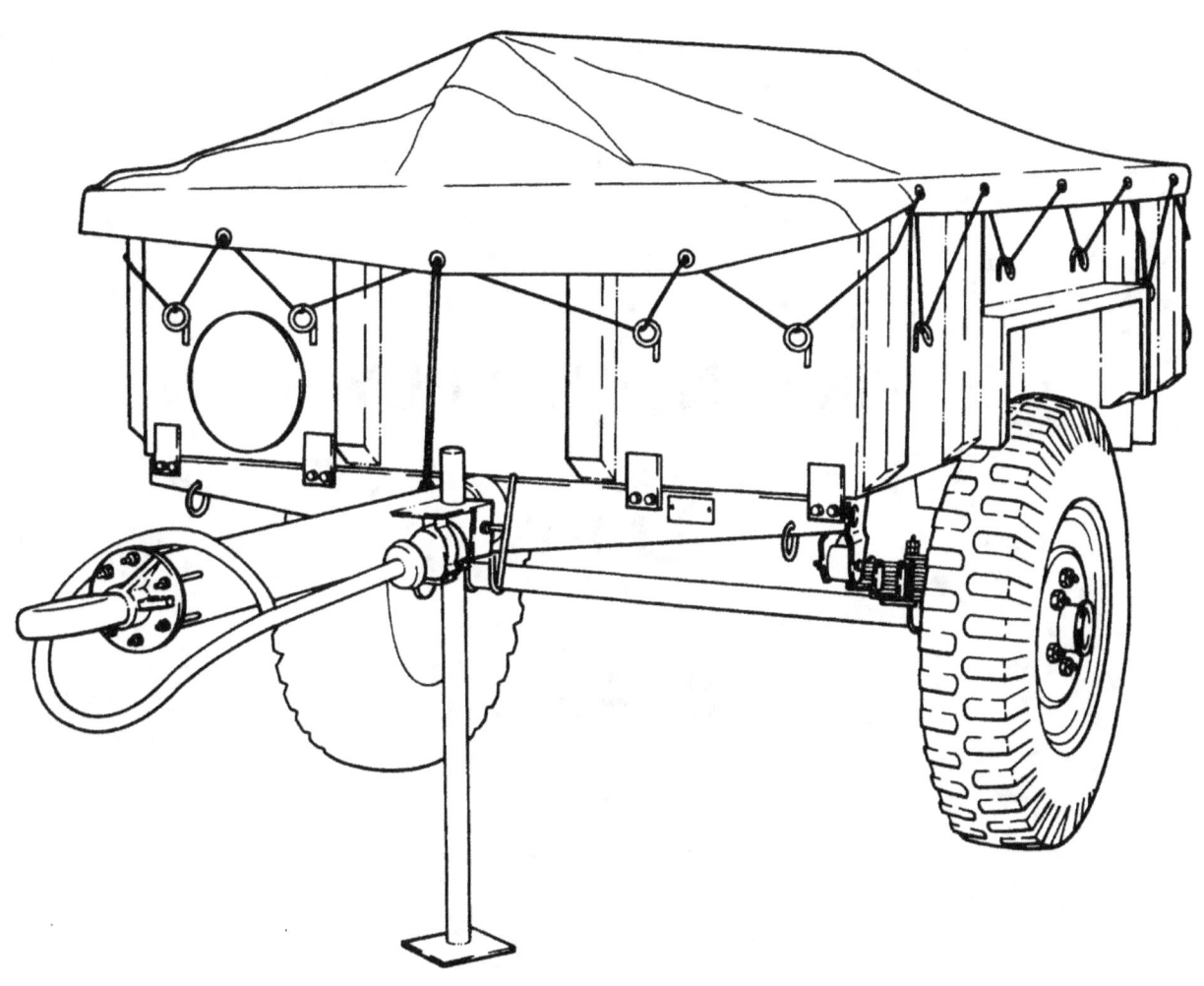

TRAILER, CARGO, ½ TON, 2 WHEELED, AUST NO 5

INTENTIONALLY LEFT BLANK

CHAPTER 2

SECTION 1

REPAIR PARTS SCALE
TEXT AND ILLUSTRATIONS

GROUP AA 02060

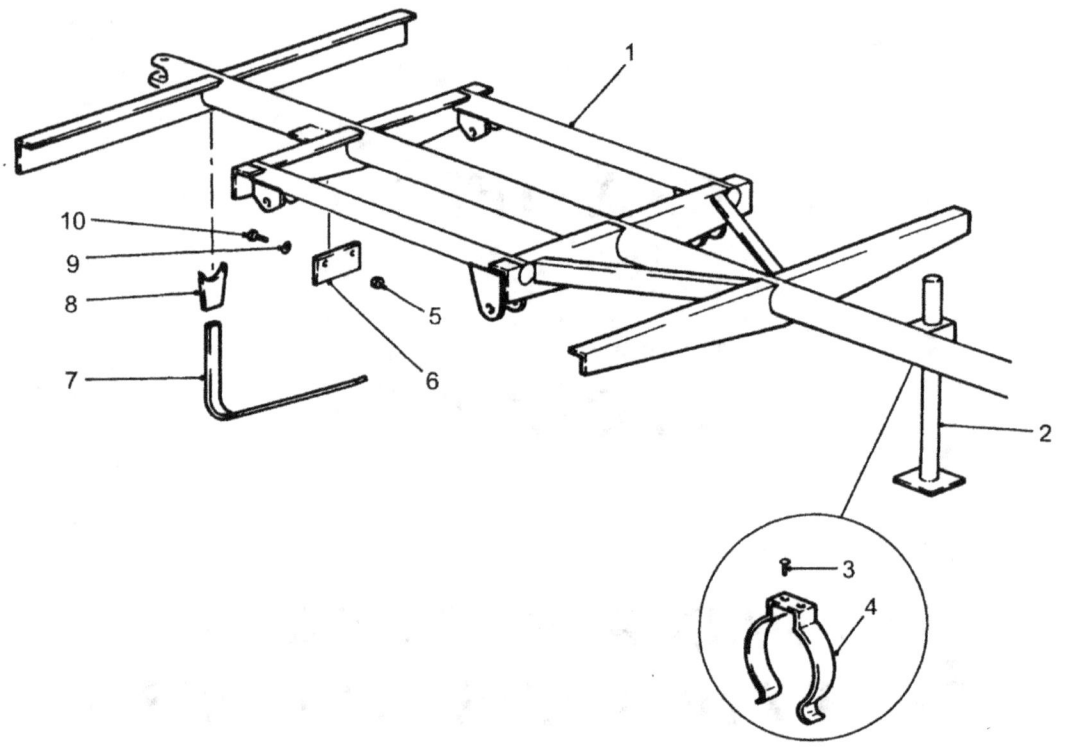

FRAME ASSEMBLY

GROUP AA
TITLE FRAME ASSEMBLY

ITEM NO	DESIGNATION	NSN MANUFACTURER CODE/PART NO SUPPLIER CODE/PART NO	NO OFF	UOI	EXP	L* M* H*
001	FRAME ASSEMBLY	Z0020/ADE(V)50-1 Z0020/ADE(V)50-1	1			
002	STAND (REFER GROUP AD)	Z0020/ADE(V)50-27 Z0020/ADE(V)50-27	1			
003	RIVET, SOLID STEEL, RD HD, 1/8 IN.DIA, 1/2 IN.LG	5320 66-013-3391 Z0544/B118 Z0544/B118	2	LB	X	LMH
004	CLIP, SPRING TENSION STEEL, 1 IN. W BY 1/16 IN. THK MATERIAL, 2-11/32 IN. DIA CAPACITY, 2 MTG HOLES	5340 66-019-0931 Z2813/ADE(V)50-35 Z4732/ADE(V)50.35	1	EA	X	LMH
005	NUT, PLAIN, HEXAGON UNF, 2B, STEEL, FORMED, ZINC COATED, 1/4 IN	5310 66-019-3956 U1030/NH604041 Z5394/NH604041L	10	EA	X	LMH
006	RUBBER FLAP, 9 IN BY 9 IN (MANUFACTURE FROM AA 901)	/NIL /NIL	1			
007	BRACKET ANGLE (MANUFACTURE FROM AA 903)	/NIL /NIL	1			
008	BRACKET (MANUFACTURE FROM AA 902)	/NIL /NIL	1			
009	WASHER, LOCK STEEL, SPLIT HELICAL-RH, CAD PLATED, 9/32 IN. ID, 15/32 IN. OD, 11/16 IN. THK	5310 66-013-0067 Z4A36/20278 14MDSW Z4A36/20278 14MDSW	6	EA	X	LMH
010	SCREW, HEXAGON HEAD 2A, SAE GRADE 5 STEEL, HEX HD, ZINC COATED, 1/4 IN.BY 1 IN.LG	5305 66-019-3656 80204/ANSI B18-2 80204/ANSI B18-2	2	EA	X	LMH
901	RUBBER SHEET, SOLID NATURAL RUBBER, CLOTH INSERT, 2 PLY, 4.5 MM THK BY 1200 MM W BY 10 M LG (USE AS REQUIRED)	9320 66-147-0619 Z0405/1/NAT4.5-3 Z0405/1/NAT4.5-3	1	MR	X	LMH
902	BAR, METAL CARBON STEEL, STRUCTURAL, GRADE 250, 6 MM THK, 250 MM W (USE AS REQUIRED)	9510 66-097-0434 Z0544/AS 1204 Z0544/AS 1204	6	KG	X	LMH

ISSUE NO: 4   REPAIR PARTS SCALE: 02060   PAGE NO: 1

* GRADE OF REPAIR (LIGHT, MEDIUM OR HEAVY)

GROUP AA 02060

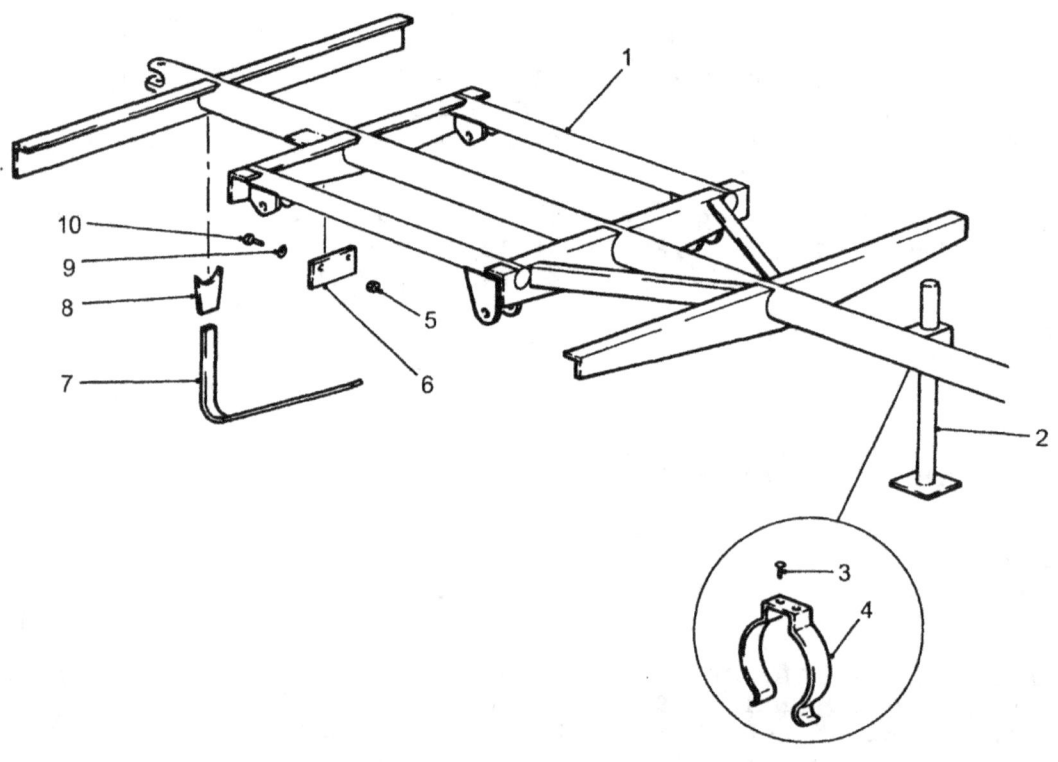

FRAME ASSEMBLY

GROUP AA
TITLE FRAME ASSEMBLY

ITEM NO	DESIGNATION	NSN MANUFACTURER CODE/PART NO SUPPLIER CODE/PART NO	NO OFF	U O I	E X P	L* M* H*
903	BAR, METAL STEEL CARBON, HOT ROLLED, GRADE 250 , FLAT, 25 MM W BY 12 MM THK (USE AS REQUIRED)	9510 66-122-8449 Z0544/AS 1204 Z0544/AS 1204	30	MR	X	LMH

ISSUE NO: 4   REPAIR PARTS SCALE: 02060   PAGE NO: 2

* GRADE OF REPAIR (LIGHT, MEDIUM OR HEAVY)

GROUP AB 02060

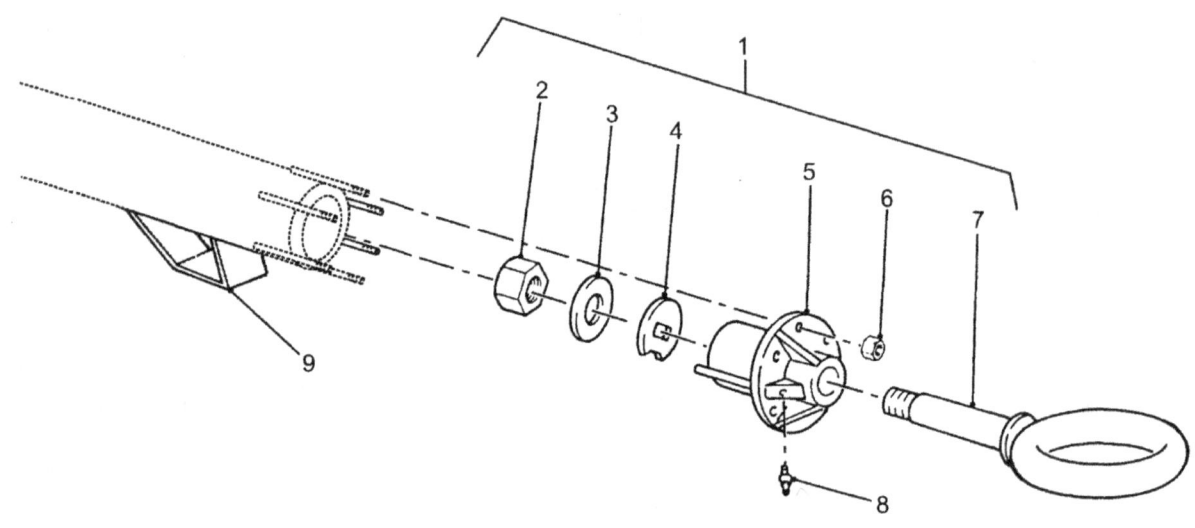

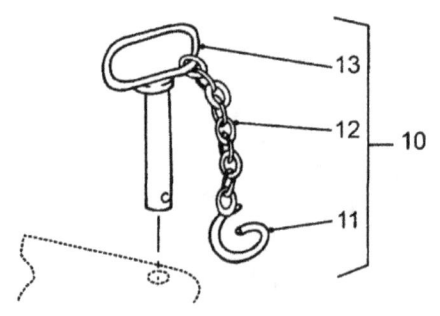

TOWING ATTACHMENT FRONT AND REAR

GROUP AB
TITLE TOWING ATTACHMANT FRONT AND REAR

ITEM NO	DESIGNATION	NSN MANUFACTURER CODE/PART NO SUPPLIER CODE/PART NO	NO OFF	U O I	E X P	L* M* H*
001	COUPLER ASSEMBLY, DRAWBAR RING C/O DRAWBAR COUPLER RING, TOWING ADAPTER, NUT, COTTER PIN, PLAIN WASHER AND GREASE NIPPLE (INCLUDES AB 002 TO AB 005, AB 007 AND AB 008)	2540 66-107-8325 Z2813/ADE(V)50-22 Z0259/ADE(V)50-22	1EA	A		LMH
002	NUT, SELF-LOCKING, HEXAGON /UNF 1IN (ALSO PART OF AB 001, MODIFY VIDE EMEI VEH H 027-13 ISSUE 2)	5310 66-049-4073 Z0635/NTD326-15-241 Z0635/NTD326-15-241	1EA		X	MH
003	WASHER, FLAT RD, STEEL, ZINC COATED, SMALL, 1 IN.BOLT SIZE, 1-7/8 IN OD BY 0.160 IN.THK (ALSO PART OF AB 001, MODIFY VIDE EMEI VEH H 027-13 ISSUE 2)	5310 66-016-8932 Z0544/AS B117 Z0544/AS B117	6EA		X	MH
004	WASHER, FLAT RD, STEEL, 37.0 MM ID, 32.0 MM WIDTH ACROSS FLATS, 75.0 MM OD, 6.0 MM THK (ALSO PART OF AB 001, MODIFY VIDE EMEI VEH H 027-13 ISSUE 2)	5310 66-132-7969 Z5454/MEA 1475 ZA229/MEA 1475	1EA		X	MH
005	ADAPTER, TOWING CI, 5-5/16 IN. DIA, 3-5/8 IN. W, 1-1/2 IN. DIA CENTRE HOLE (ALSO PART OF AB 001)	2540 66-016-9984 Z2813/ADE(V)50-24 Z2813/ADE(V)50-24	1EA		X	MH
006	NUT, SELF-LOCKING, HEXAGON UNF 2B STEEL CAD PLATED W/CHROMATE 3/8 IN BOLT SIZE ALL METAL TYPE (ALSO PART OF AB 001)	5310 00-959-1488 Z0365/NPD126/11/6 Z0365/NPD126/11/6	6EA		X	MH
007	COUPLER, DRAWBAR, RING 3 IN. ID, 6 IN. OD, 4-11/16 16 IN. LG SHANK (ALSO PART OF AB 001)	2540 66-017-0060 Z2813/ADE(V)50-23 Z2813/ADE(V)50-23	1EA		X	MH
008	FITTING, LUBRICATION HYDRAULIC TYPE, STEEL, STRAIGHT, CAD PLATED, 1/8 IN.BSP W/CHECK VALVE, 7/32 IN.STEM LG, 13/16 IN.O/A LG (ALSO PART OF AB 001)	4730 66-038-0295 Z0200/H29 Z2966/H29	1EA		X	LMH

ISSUE NO: 4   REPAIR PARTS SCALE: 02060   PAGE NO: 3

* GRADE OF REPAIR (LIGHT, MEDIUM OR HEAVY)

GROUP AB 02060

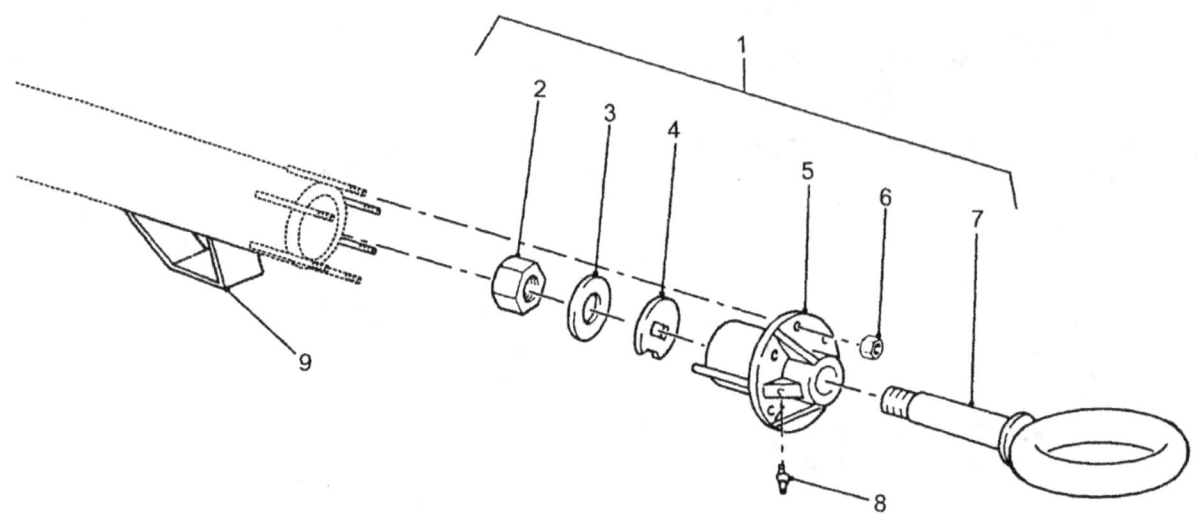

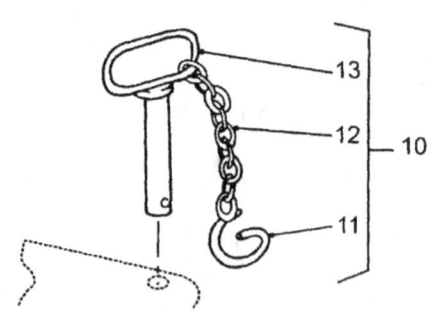

TOWING ATTACHMENT FRONT AND REAR

GROUP AB
TITLE TOWING ATTACHMANT FRONT AND REAR

ITEM NO	DESIGNATION	NSN MANUFACTURER CODE/PART NO SUPPLIER CODE/PART NO	NO OFF	U O I	E X P	L* M* H*
009	BRACKET, STAND (MANUFACTURE FROM AB 901)	/NIL /NIL	1			
010	PIN AND HOOK ASSEMBLY REAR TOWING (INCLUDES AB 011 TO AB 013)	2540 66-017-5150 Z2813/ADE(V)50-6 Z2813/ADE(V)50-6	1	EA	X	LMH
011	HOOK (ALSO PART OF AB 010)	Z0020/ADE(V)50-8 Z0020/ADE(V)50-8	1			
012	CHAIN, WELDLESS STEEL, ZINC COATED, DOUBLE JACK STYLE , 14 GAUGE (MANUFACTURE FROM AB 902, ALSO PART OF AB 010)	/NIL /NIL	1			
013	PIN (ALSO PART OF AB 010)	Z0020/ADE(V)50-7 Z0020/ADE(V)50-7	1			
901	BAR, METAL CARBON STEEL, HOT ROLLED, GRADE 250 , FLAT, 50 MM W BY 8 MM THK (USE AS REQUIRED)	9510 66-097-0964 Z0544/AS 3678 Z0544/AS 3678	3	MR	X	MH
902	CHAIN, WELDLESS STEEL, ZINC COATED, DOUBLE JACK STYLE , 14 GAUGE (USE REQUIRED)	4010 66-044-0246 Z0549/112 X 14G Z0549/112 X 14G	1	MR	X	LMH

ISSUE NO: 4   REPAIR PARTS SCALE: 02060   PAGE NO: 4

* GRADE OF REPAIR (LIGHT, MEDIUM OR HEAVY)

GROUP AD 02060

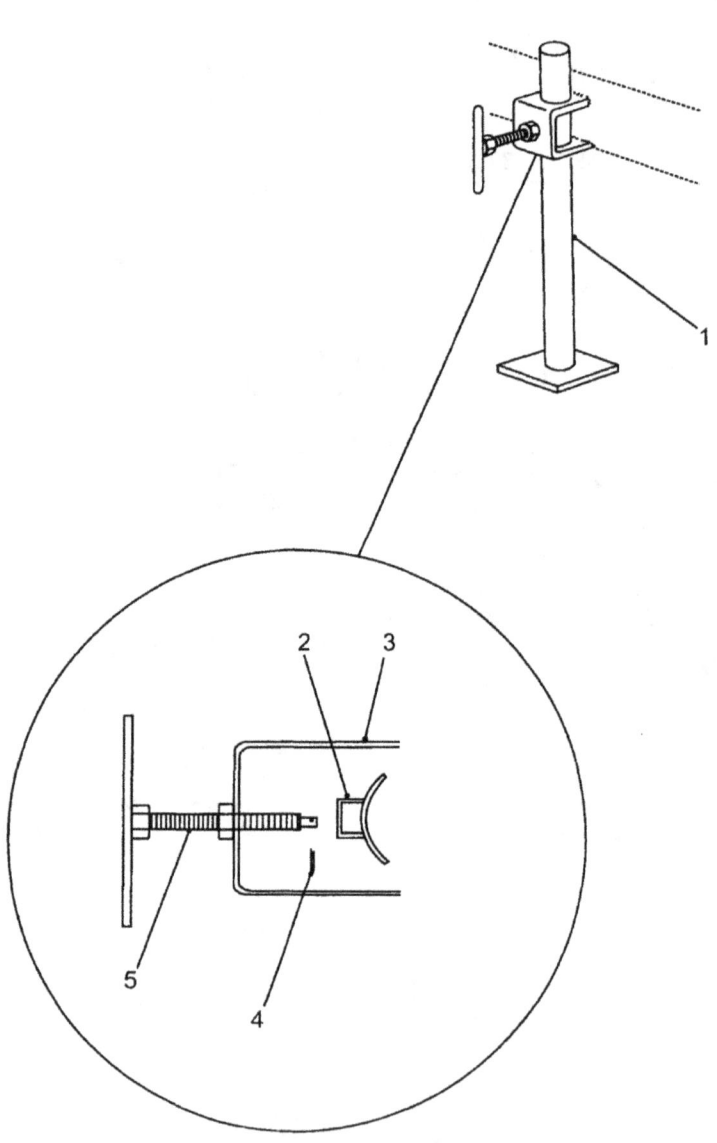

STAND LEVELLING

GROUP AD
TITLE STAND LEVELLING

ITEM NO	DESIGNATION	NSN MANUFACTURER CODE/PART NO SUPPLIER CODE/PART NO	NO OFF	U O I	E X P	L* M* H*
001	SUPPORT, LEVELLING, VEHICLE	2590 66-017-5149 Z2813/ADE(V)50-27 Z2813/ADE(V)50-27	1	EA	X	LMH
002	SHOE, LEVELLING SUPPORT	2590 66-016-9986 Z2813/ADE(V)50-29 Z2813/ADE(V)50-29	1	EA	X	LMH
003	BRACKET, SUPPORT BRACKET SUPPORT TRAILER LEG	Z2813/ADE(V)50-9 Z2813/ADE(V)50-9	1			
004	PIN, COTTER SPLIT, STEEL, 1.6 MM NOM DIA, 20 MM LG	5315 66-097-0325 Z0544/AS 1236 Z0544/AS 1236	1	EA	X	LMH
005	SCREW, LEVELLING SUPPORT	2590 66-016-9985 Z2813/ADE(V)50-28 Z2813/ADE(V)50-28	1	EA	X	LMH

ISSUE NO: 4   REPAIR PARTS SCALE: 02060   PAGE NO: 5

* GRADE OF REPAIR (LIGHT, MEDIUM OR HEAVY)

GROUP B 02060

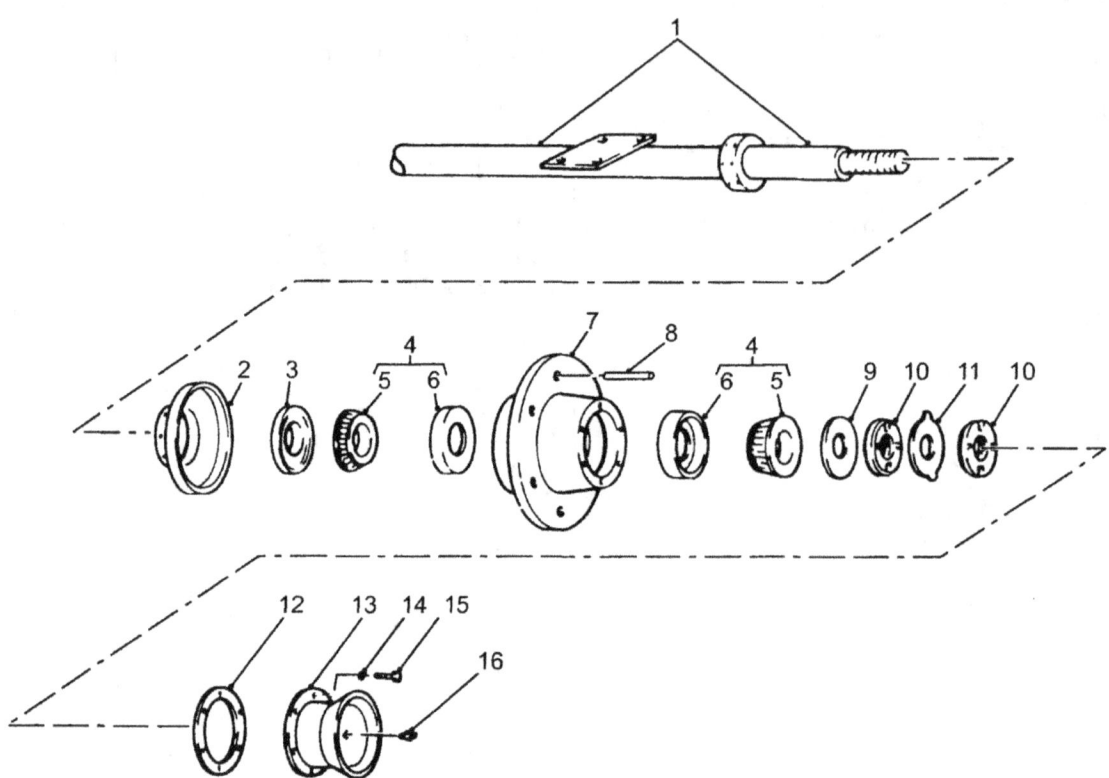

AXLE ASSEMBLY

GROUP B
TITLE AXLE ASSEMBLY

ITEM NO	DESIGNATION	NSN MANUFACTURER CODE/PART NO SUPPLIER CODE/PART NO	NO OFF	U O I	E X P	L* M* H*
001	AXLE, VEHICULAR, NONDRIVING STEEL, TUBULAR, 59 IN.LG BY 1-7/8 IN.DIA	2530 66-017-0061 Z2813/ADE(V)50-15 Z2813/ADE(V)50-15	1 EA		X	MH
002	DEFLECTOR, DIRT AND LIQUID	2530 66-016-9987 Z2813/ADE(V)50-16 Z2813/ADE(V)50-16	2 EA		X	LMH
901	SCREW, MACHINE UNF, 2A, STEEL, RD HD, SLOT DRIVE, ZINC PLATED, NO 8 BY 3/4IN LG	5305 66-036-9099 80204/ANSI B18-6-3 80204/ANSI B18-6-3	2 EA		X	LMH
902	WASHER, LOCK STEEL, SINGLE TURN-RH, ZINC PLATED, NO 10 SCREW SIZE, 3/64 IN.THK	5310 66-016-5997 K7766/BS 1802 Z0745/3-16B0LTDIASQSEC	2 EA		X	LMH
003	SEAL, PLAIN ENCASED STEEL, SYNTHETIC RUBBER ELEMENT, 2-3/8 IN.SHAFT SIZE, 3.376 IN.OD BY 1/2 IN.W	5330 66-122-8827 Z0639/50178 Z0639/50178	2 EA		X	LMH
004	BEARING, ROLLER, TAPERED COMPLETE AS ASSEMBLY OF CUP TAPERED AND CONE AND ROLLERS (INCLUDES B 005 AND B 006)	3110 00-100-3872 60038/359S354A 60038/359S354A	4 EA		X	LMH
005	CONE AND ROLLERS, TAPERED ROLLER BEARING, SINGLE ROW, RETAINER TYPE, STRAIGHT BORE, NORMAL ANGLE, 1.8125 IN BY 8.8540 IN (ALSO PART OF B 004)	60038/359S 60038/359S	4			
006	CUP, TAPERED ROLLER BEARING, SINGLE ROW BY 3.3464 IN BY 0.6875 IN (ALSO PART OF B 004)	Z0992/354A Z0992/354A	4			
007	HUB, WHEEL, VEHICULAR	2530 66-016-9988 Z2813/ADE(V)50-17 Z2813/ADE(V)50-17	2 EA		X	LMH
008	STUD, PLAIN UNF 2A/ BSF, R STEEL, CAD PLATED, 9/16 IN.BY 1-5/8 IN.LG, SCREWED 1/2 IN.AND 1 IN.LG	5307 66-017-0071 Z2813/ADE(V)50-18 ZA212/55798	10 EA		X	LMH
009	WASHER, KEY STEEL, CASEHARDENED, 1.817 IN.MAX ID , 2.500 IN.OD, 1 INT KEY	5310 00-679-4584 60038/K91509 ZW977/K91509	2 EA		X	LMH
010	NUT, PLAIN, ROUND 45 MM, 19 TPI, WHIT FORM	5310 66-122-9246 Z2813/ADE(V)50-14-9 52679/HM9	4 EA		X	LMH

ISSUE NO: 4   REPAIR PARTS SCALE: 02060   PAGE NO: 6

* GRADE OF REPAIR (LIGHT, MEDIUM OR HEAVY)

GROUP B 02060

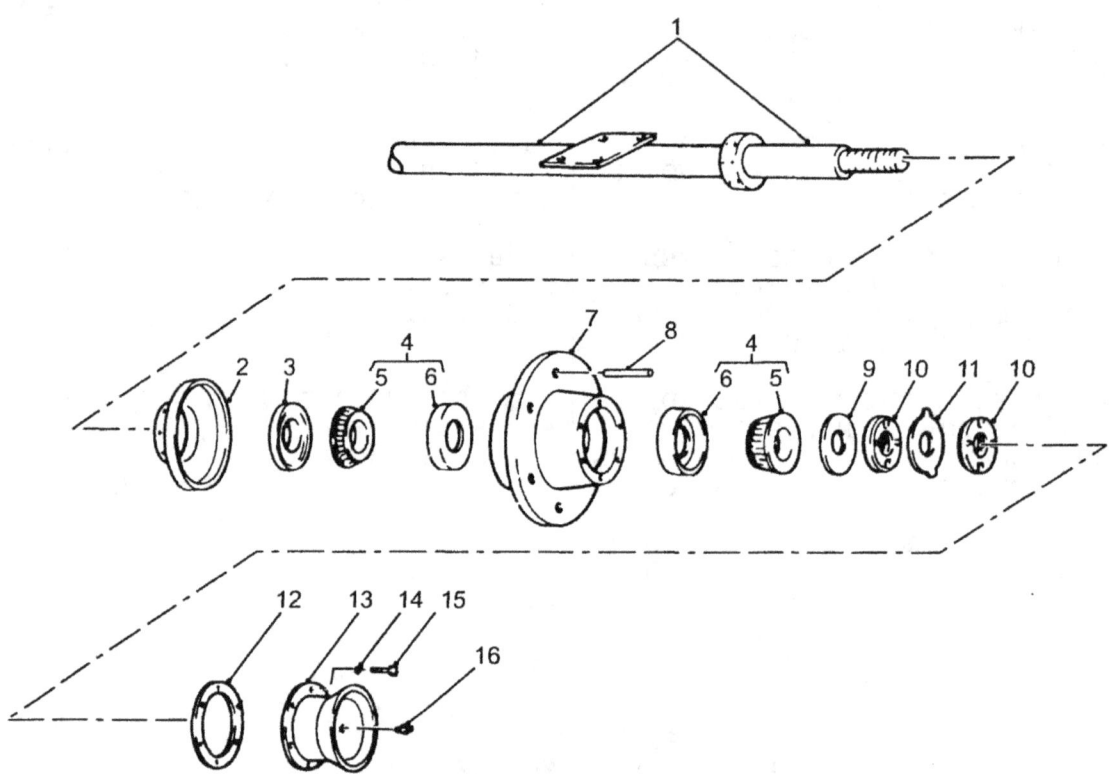

AXLE ASSEMBLY

GROUP B
TITLE AXLE ASSEMBLY

ITEM NO	DESIGNATION	NSN MANUFACTURER CODE/PART NO SUPPLIER CODE/PART NO	NO OFF	U O I	E X P	L* M* H*
011	WASHER, KEY STEEL, 2-11/16 IN.OD BY 1/16IN.LGK	5310 66-016-9990 Z2813/ADE(V)50-14/10 52676/MB9	2	EA	X	LMH
012	GASKET AXLE GREASE CAP	5330 66-036-9096 Z2813/ADE(V)50-36 Z2813/ADE(V)50-36	2	EA	X	LMH
013	CAP, GREASE	2530 66-017-0072 Z2813/ADE(V)50-19 Z2813/ADE(V)50-19	2	EA	X	LMH
014	WASHER, LOCK STEEL, SPLIT HELICAL-RH, CAD PLATED & PASSIVATED, 5/16 IN. ID, 19/32 IN. OD, 11/16 IN. THK	5310 66-013-0070 K7766/BS 1802 Z4A36/20282 516MDSW	12	EA	X	LMH
015	BOLT, MACHINE UNC, 2A, SAE GRADE 5 STEEL, HEX HD, ZINC COATED, 5/16 IN.BY 3/4 IN.LG	5306 66-019-3719 80204/B18-2 80204/B18-2	12	EA	X	LMH
016	FITTING, LUBRICATION HYDRAULIC TYPE, STEEL, STRAIGHT, CAD PLATED, 1/8 IN.BSP W/CHECK VALVE, 7/32 IN.STEM LG, 13/16 IN.O/A LG	4730 66-038-0295 Z0200/H29 Z2966/H29	2	EA	X	LMH

ISSUE NO: 4 REPAIR PARTS SCALE: 02060 PAGE NO: 7

* GRADE OF REPAIR (LIGHT, MEDIUM OR HEAVY)

GROUP D

02060

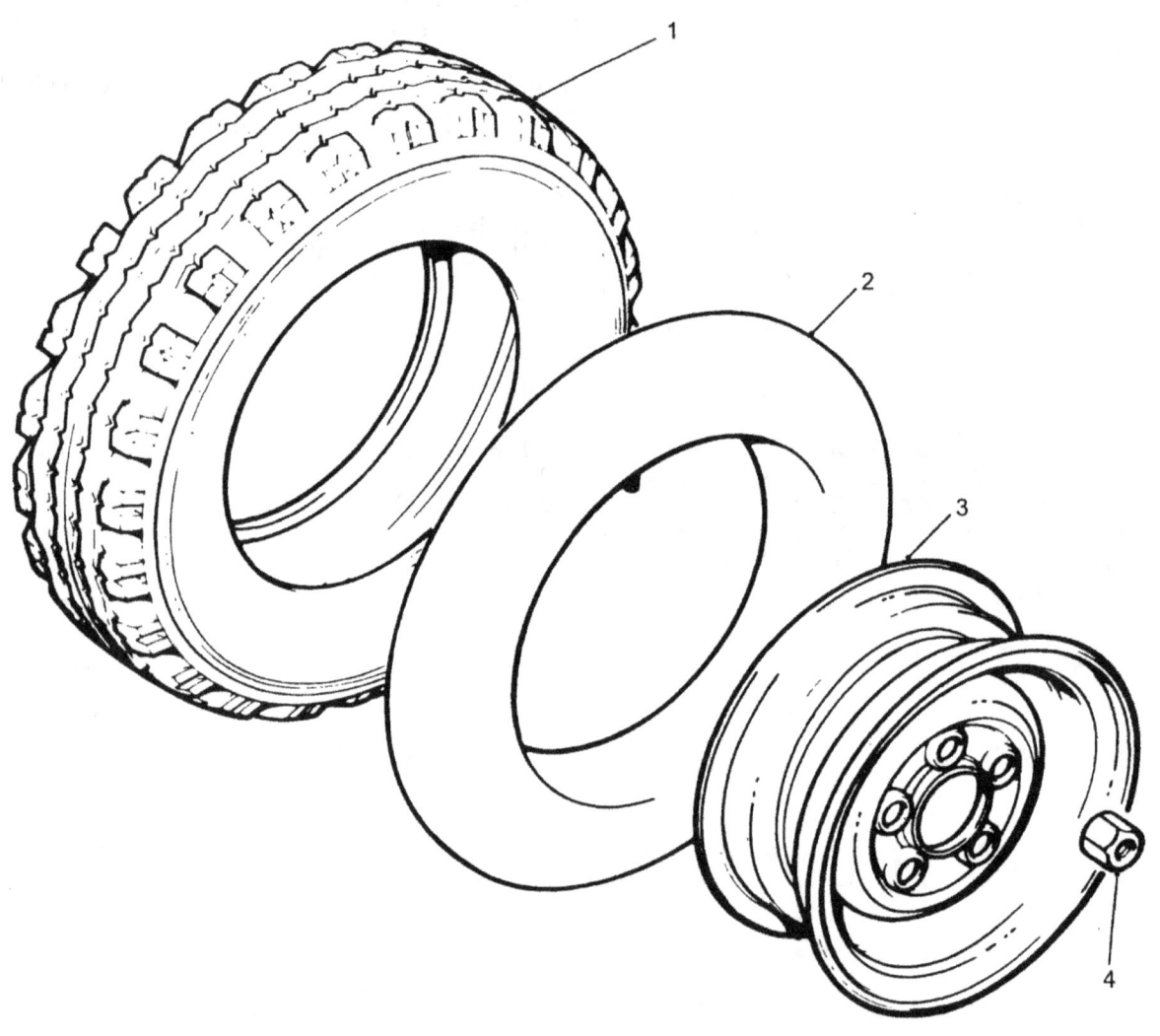

WHEEL, RIM AND TYRE

GROUP D
TITLE WHEEL, RIM AND TYRE

ITEM NO	DESIGNATION	NSN / MANUFACTURER CODE/PART NO / SUPPLIER CODE/PART NO	NO OFF	U O I	E X P	L* M* H*
001	TIRE, PNEUMATIC, VEHICULAR 7.50R16LT/10, RADIAL PLY, STEEL TREK	2610 66-108-0508 Z3551/2233346 Z3551/2233346	2EA	A		LMH
002	INNER TUBE, PNEUMATIC TIRE, VEHICULAR 7.50-16, TR15, OFF CENTRE, LIGHT TRUCK	2610 66-010-7864 Z3551/2822802 Z3551/2822802	2EA		X	LMH
003	WHEEL, PNEUMATIC TYRE	2530 99-406-7607 U1030/272309 Z5394/R272309	2EA	A		
004	NUT, PLAIN, CONE SEAT, HEXAGON BSF, STEELPHOSPHATED, 0.562 IN., 0.920 IN.W A/F, 1.000 IN.H	5310 66-010-7012 U1030/217361 Z5394/217361	10EA		X	LMH

ISSUE NO: 4   REPAIR PARTS SCALE: 02060   PAGE NO: 8

* GRADE OF REPAIR (LIGHT, MEDIUM OR HEAVY)

GROUP EA 02060

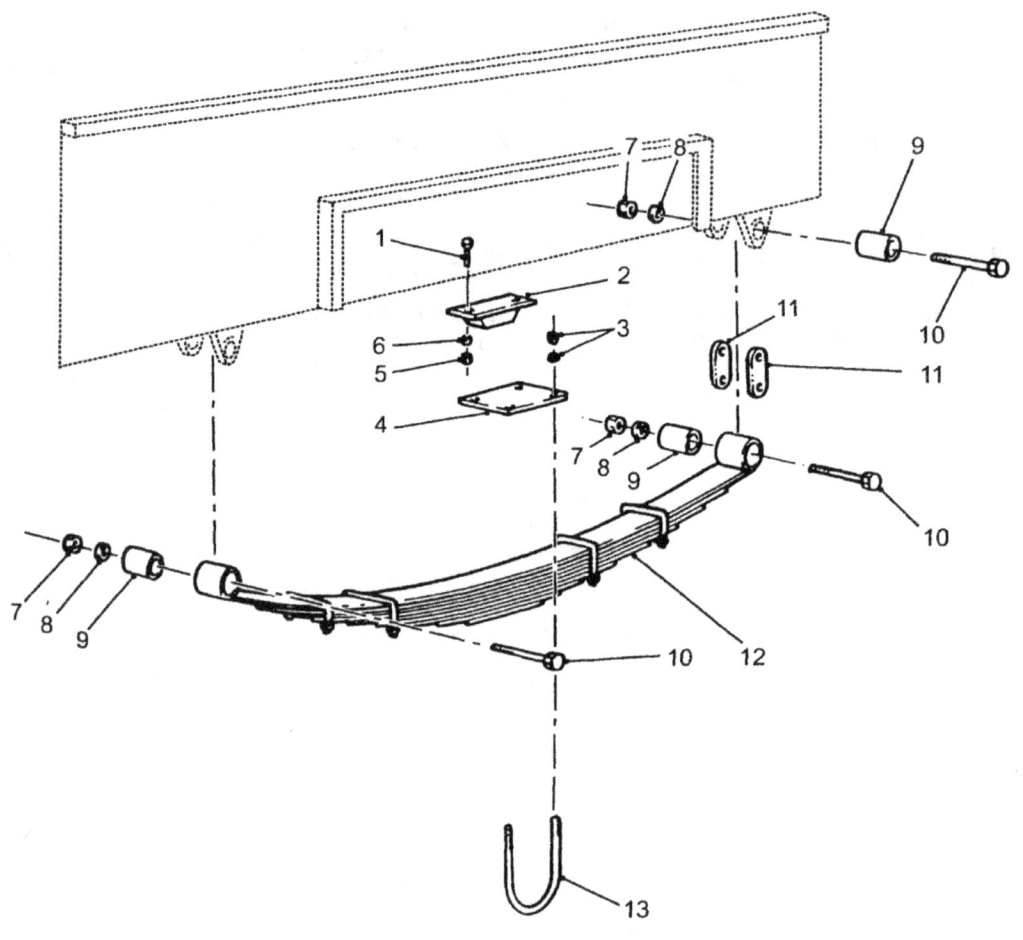

SPRINGS AND MOUNTINGS

GROUP EA
TITLE SPRINGS AND MOUNTINGS

ITEM NO	DESIGNATION	NSN MANUFACTURER CODE/PART NO SUPPLIER CODE/PART NO	NO OFF	U O I	E X P	L* M* H*
001	BOLT, MACHINE UNF, 2A, SAE GRADE 5 STEEL, HEX HD, ZINC COATED, 5/16IN BY 1-1/8IN LG	5306 66-019-3767 Z0544/AS 2465 80204/B18-2	4	EA	X	LMH
002	BUMPER, NONMETALLIC RUBBER, TAPERED, 2-7/8 IN.H O/A BY 5 IN.LG O/A	5340 66-010-9530 Z2813/TSE(V)54-202 Z2813/TSE(V)54-202	4	EA	X	LMH
003	NUT, PLAIN, HEXAGON UNF, 2B, STEEL, ZINC PLATED, 7/16 IN	5310 66-019-3959 Z2C66/3125 96906/MS35691-29	16	EA	X	LMH
004	PLATE	 Z0020/ADE(V)50-31 Z0020/ADE(V)50-31	2			
005	NUT, PLAIN, HEXAGON UNF, 2B, STEEL, GRADE 0, LOCK, ZINC COATED, 5/16IN	5310 66-010-7415 Z0544/AS 2465 ZA212/TW44536	4	EA	X	LMH
006	WASHER, LOCK STEEL, CADMIUM PLATED, 5/16 IN.BOLT SIZE, SINGLE TURN SPRING WASHER	5310 66-016-5980 K7766/BS 1802 Z5394/WM600051	4	EA	X	LMH
007	NUT, PLAIN, HEXAGON UNF, 2B, STEEL, ZINC PLATED, 1/2 IN	5310 66-019-3960 80204/B18-2-2 80204/B18-2-2	6	EA	X	LMH
008	WASHER, LOCK STEEL, CADMIUM PLATED, 1/2 IN.BOLT SIZE, SINGLE TURN SPRING WASHER	5310 66-016-5993 K7766/BS 1802 K7766/BS 1802	6	EA	X	LMH
009	MOUNT, RESILIENT, GENERAL PURPOSE 2IN.LG BY 1.062IN.OD BY 0.500IN.ID	5340 66-017-5340 Z0366/E20822 Z0273/S19	6	EA	X	LMH
010	BOLT, MACHINE UNF, 2A, SAE GRADE 5 STEEL, HEX HD, ZINC COATED, 1/2IN BY 4IN LG	5306 66-019-3897 Z0544/AS 2465 80204/B18-2	6	EA	X	LMH
011	CONNECTING LINK, RIGID STEEL, 4.500 IN.LG BY 1.750 IN.W BY 0.312 IN.THK, W/TWO 0.500 IN.HOLES	3040 66-017-5338 Z2813/ADE(V)50-30 Z2813/ADE(V)50-30	4	EA	X	LMH
012	SPRING ASSEMBLY, LEAF STEEL, 10 LEAVES, 36.000 IN.LG BY 2.750 IN.THK BY 1.750 IN.LG	2510 66-021-0191 Z2813/ADE(V)50-20/1 Z2813/ADE(V)50-20/1	2	AY	X	MH
901	BOLT, MACHINE UNF, 2A, STEEL, CHEESE HD, 5/16IN.BY 3-1/4IN (CENTRE BOLT)	5306 66-017-0073 Z2813/ADE(V)50-20-8 Z2813/ADE(V)50-20-8	1	EA	X	LMH

ISSUE NO: 4   REPAIR PARTS SCALE: 02060   PAGE NO: 9

* GRADE OF REPAIR (LIGHT, MEDIUM OR HEAVY)

GROUP EA 02060

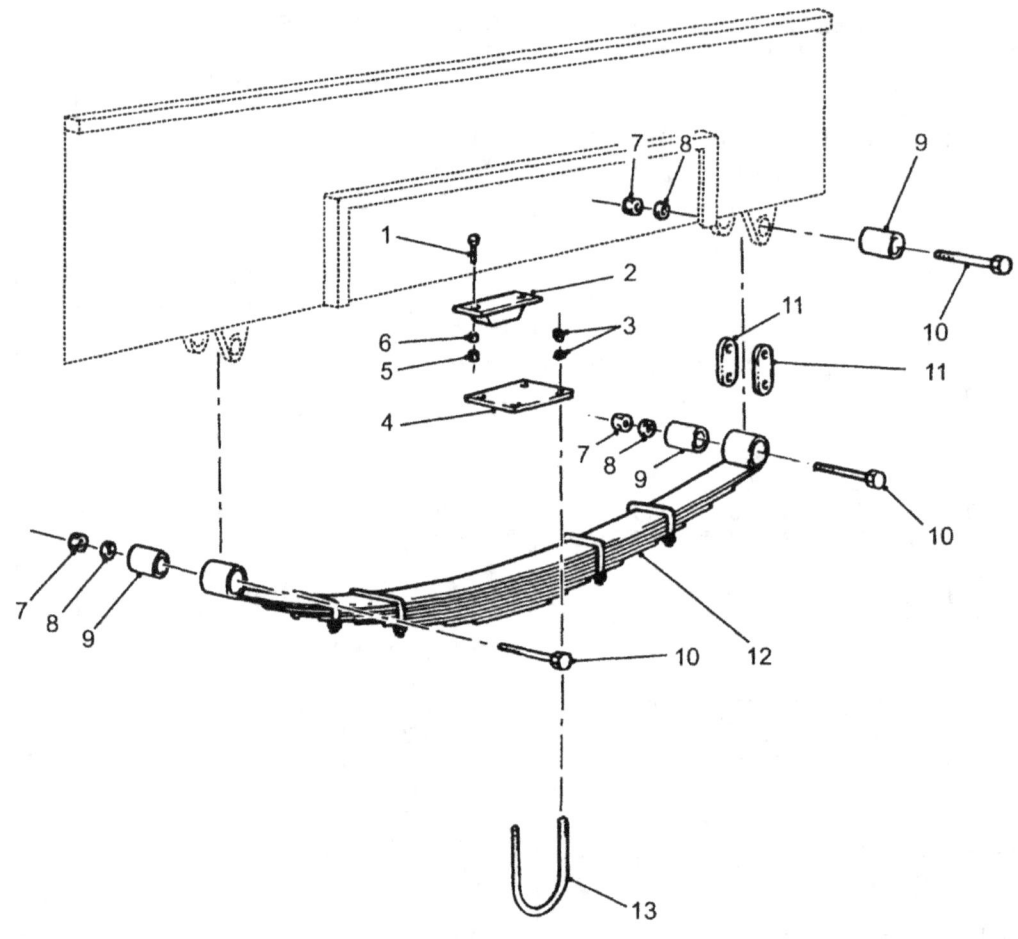

SPRINGS AND MOUNTINGS

GROUP EA
TITLE SPRINGS AND MOUNTINGS

ITEM NO	DESIGNATION	NSN MANUFACTURER CODE/PART NO SUPPLIER CODE/PART NO	NO OFF	U O I	E X P	L* M* H*
902	WASHER, LOCK STEEL, SPLIT HELICAL-RH, CAD PLATED & PASSIVATED, 5/16 IN. ID, 19/32 IN. OD, 11/16 IN. THK (USE WITH EA 901)	5310 66-013-0070 Z4A36/20282 516MDSW 17600/DX252	1 EA		X	LMH
903	NUT, PLAIN, HEXAGON UNF, 2B, STEEL, FORMED, ZINC COATED, 5/16 IN (USE WITH EA 901)	5310 66-019-3957 80204/ANSI B18-2-2 Z1272/034-079	1 EA		X	LMH
904	ALIGNMENT CLIP, LEAF SPRING	2510 66-036-6132 Z2813/ADE(V)50-21-2 Z2813/ADE(V)50-21-2	2 EA		X	MH
905	ALIGNMENT CLIP, LEAF SPRING	2510 66-036-6131 Z2813/ADE(V)50-21-1 Z2813/ADE(V)50-21-1	2 EA		X	MH
906	SPACER, SLEEVE STEEL, 0.438IN ID, 0.566IN OD BY 1.750IN LG O/A	5365 66-036-6133 Z2813/ADE(V)50-20-6 Z2813/ADE(V)50-20-6	4 EA		X	MH
907	BOLT, MACHINE UNF, 2A, SAE GRADE 5 STEEL, HEX HD, ZINC COATED, 1/4IN BY 2-1/2IN LG	5306 66-019-3763 Z0544/AS 2465 80204/B18-2	4 EA		X	MH
908	WASHER, LOCK STEEL, SPLIT HELICAL-RH, CAD PLATED, 9/32 IN. ID, 15/32 IN. OD, 11/16 IN. THK	5310 66-013-0067 Z4A36/20278 14MDSW Z4A36/20278 14MDSW	4 EA		X	MH
909	NUT, PLAIN, HEXAGON UNF, 2B, STEEL, FORMED, ZINC COATED, 1/4 IN	5310 66-019-3956 U1030/NH604041 Z5394/NH604041L	4 EA		X	MH
013	BOLT, U UNF, 2A, STEEL, 7/16IN.BY 6-1/2IN.LG , 1-7/8IN.W BETWEEN SHANKS	5306 66-017-0074 Z2813/ADE(V)50-32 Z2813/ADE(V)50-32	4 EA		X	LMH

ISSUE NO: 4   REPAIR PARTS SCALE: 02060

* GRADE OF REPAIR (LIGHT, MEDIUM OR HEAVY)

GROUP FA 02060

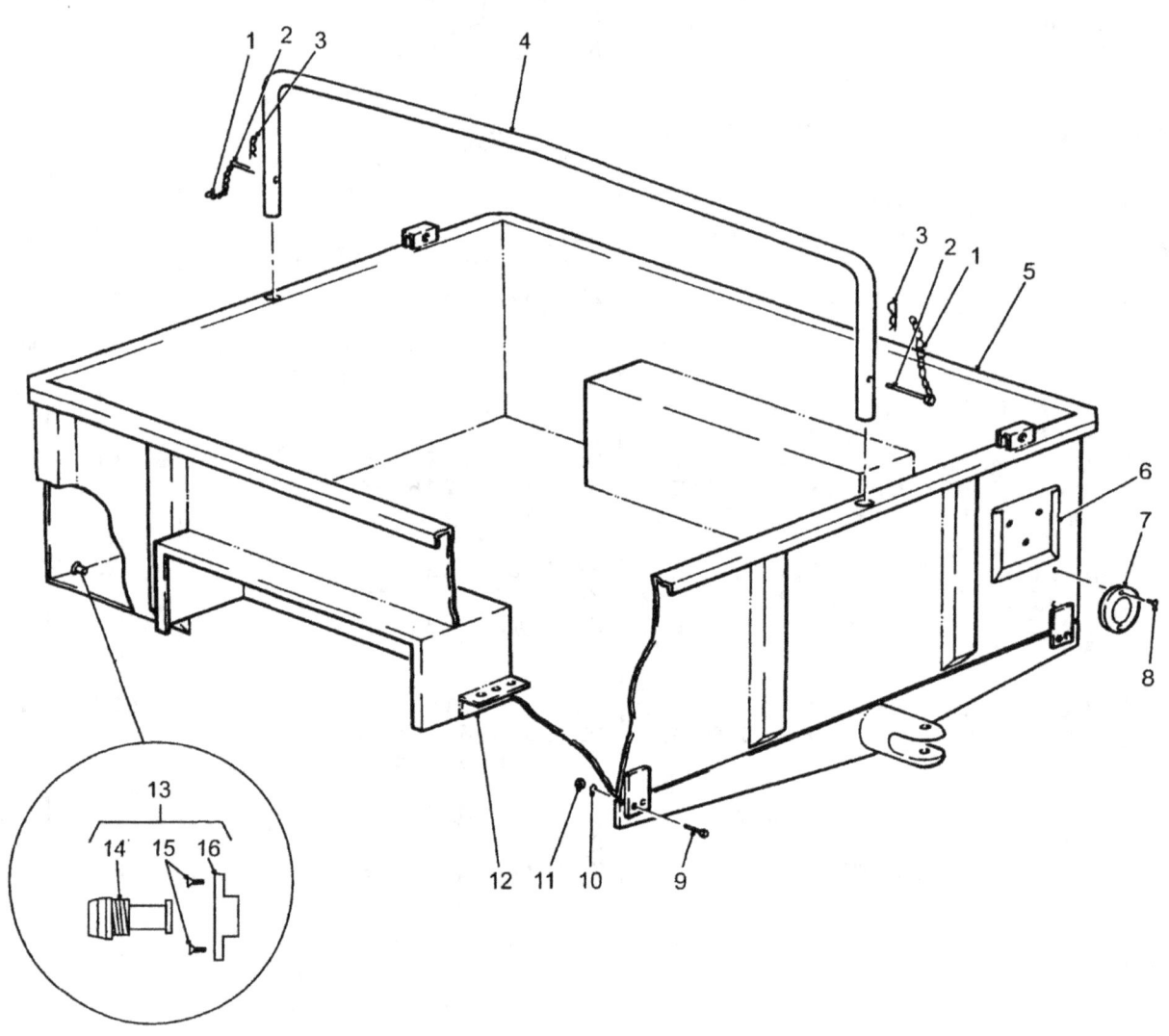

CARGO BODY AND MOUNTINGS

GROUP   FA
TITLE   CARGO BODY AND MOUNTINGS

ITEM NO	DESIGNATION	NSN / MANUFACTURER CODE/PART NO / SUPPLIER CODE/PART NO	NO OFF	U O I	E X P	L* M* H*
001	CHAIN, WELDLESS STEEL, ZINC COATED, DOUBLE JACK STYLE, 14 GAUGE (MANUFACTURE FROM AB 902)	/NIL /NIL	1			
002	PIN, STRAIGHT, HEADED STEEL, 3IN LG BY 5/16IN DIA, W/CHAIN LINK HD	5315 66-017-0070 Z2813/ADE(V)5-9 Z2813/ADE(V)5-9	2	EA	X	LMH
003	PIN, LOCK STEEL, 1-13/16IN LG BY 14 SWG	5315 66-017-5137 Z2813/ADE(V)5-10 Z2813/ADE(V)5-10	2	EA	X	LMH
004	BOW	Z0020/ADE(V)51-2 Z0020/ADE(V)51-2	1			
005	BODY CARGO	Z0020/ADE(V)53-2 Z0020/ADE(V)53-2	1			
006	HOLDER, VEHICLE UNIT AND FORMATION SIGN PLATE STEEL, PAINTED DEEP BRONZE GREEN, 6.094 IN. LG, 0.297 IN. O/A W, 6.000 IN. H, W/3 RAISED MTG HOLES, TO SUIT	2590 66-016-0815 Z2813/ADE(V)43-1 Z5394/HYK2674	2	EA	X	LMH
007	REFLECTOR, INDICATING, CLEARANCE PLASTIC, RED, 3 IN.DIA LENS, 4-1/2 IN.O/A DIA	9905 66-010-9168 Z0083/96 ZR910/96	2	EA	X	LMH
008	SCREW, TAPPING THREAD FORMING, STEEL, PAN HD, SLOTTED, DOG TYPE POINT, NO 10 BY 5/16 IN	5305 66-013-9530 80204/B18-6-4 ZA212/B18-6-4	4	EA	X	LMH
009	SCREW, HEXAGON HEAD 2A, STEEL, HEX HD, ZINC PLATED, 3/8 IN.BY 1 IN.LG	5305 66-019-3680 80204/B18-2-1 80204/B18-2-1	24	EA	X	MH
010	WASHER, LOCK SPRING, STEEL, SINGLE TURN, SQ SECTION, CAD PLATED, 3/8 IN BOLT SIZE, 0.590 IN OD, 0.092 IN THK	5310 66-016-5995 K7766/BS 1802 Z2C66/3751	24	EA	X	MH
011	NUT, PLAIN, HEXAGON UNF, 2B, STEEL, GRADE 0, LOCK, ZINC COATED, 3/8IN	5310 66-011-0004 ZA212/AS 2465 ZA212/TW36017	24	EA	X	MH
012	BRACKET ASSEMBLY BRACKET ASSEMBLY, HOLD DOWN	2510 66-024-5131 Z2813/ADE(V)53-1 Z2813/ADE(V)53-1	4	AY	X	MH

ISSUE NO: 4   REPAIR PARTS SCALE: 02060

* GRADE OF REPAIR (LIGHT, MEDIUM OR HEAVY)

GROUP FA 02060

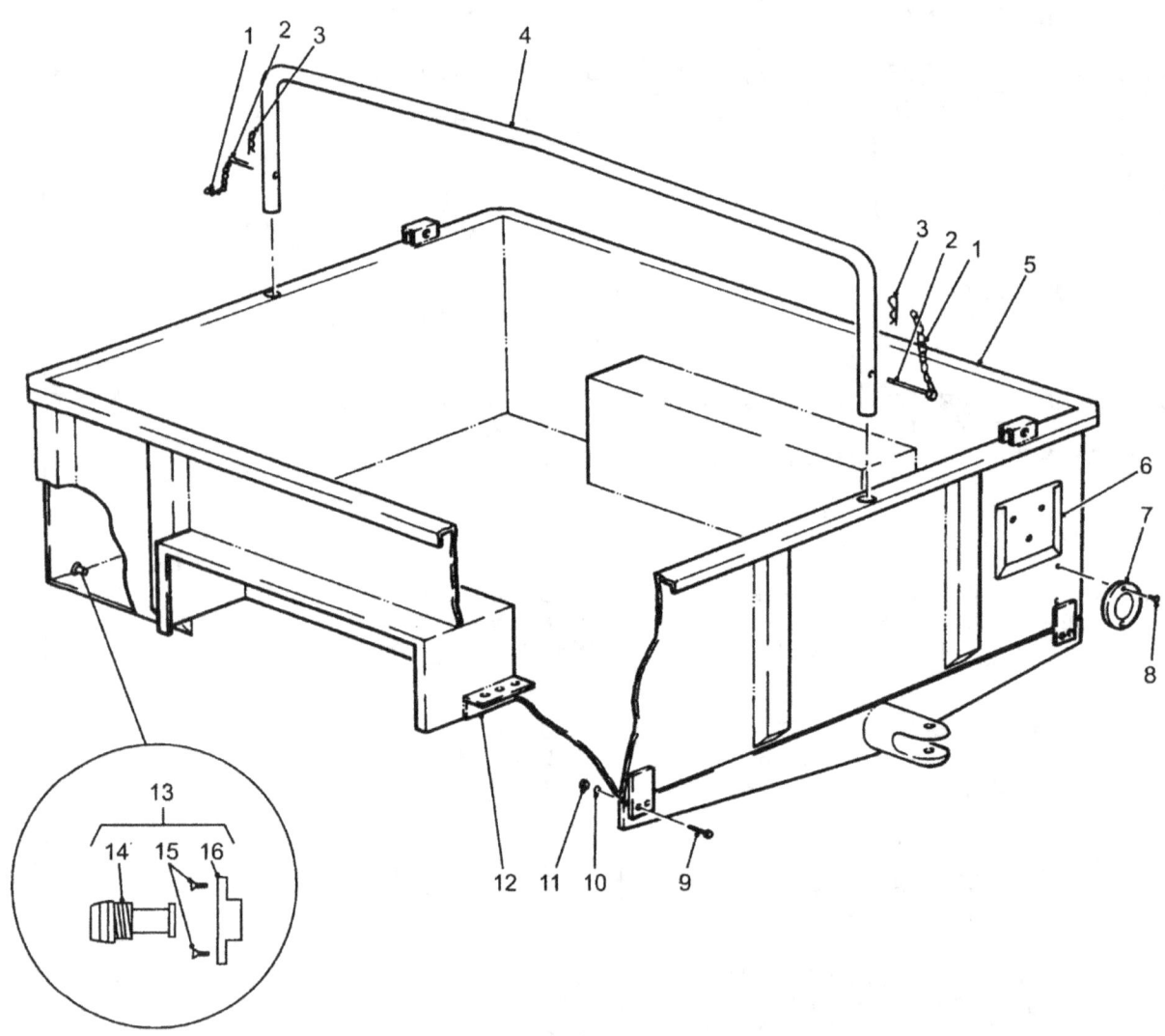

CARGO BODY AND MOUNTINGS

GROUP FA
TITLE CARGO BODY AND MOUNTINGS

ITEM NO	DESIGNATION	NSN MANUFACTURER CODE/PART NO SUPPLIER CODE/PART NO	NO OFF	UOI	EXP	L* M* H*
013	PLUG ASSY, DRAIN TRANSOM PLUG DRAIN, BRONZE BODY C/W PLASTIC BUNG (INCLUDES FA 014 TO FA 016)	2040 66-068-5743 Z2767/RF737 Z4311/270-75	2	EA	X	LMH
014	PLUG, DRAIN, TRANSOM BUNG, PLASTIC (ALSO PART OF FA 013)	2040 66-060-0704 Z2767/RF738 Z4311/270-76	2	EA	X	LMH
015	SCREW, MACHINE BSW, FREE FIT, BRASS, FLAT CSK HD, 3/16IN BY 1IN (ALSO PART OF FA 013)	5305 66-012-3974 Z0544/AS B50 Z0544/AS B50	4	EA	X	LMH
016	BODY BRASS (ALSO PART OF FA 013)	Z1785/PN592 Z1785/PN592	2			
901	PLUG, PIPE BSP, MI, SOLID HEAD AND BODY, ZINC COATED, 1-1/2 IN.THD SIZE	4730 66-012-8299 K7766/BS 143 K7766/BS 143	1	EA	X	LMH
902	WIRE, NONELECTRICAL STEEL, CARBON, SOFT, 2.5 MM DIA, BRIGHT, 5.5 KN, FENCING WIRE	9505 66-011-9348 Z0544/AS 1394 Z0544/AS 1394	6	KG	X	LMH
903	SIGN, BRIDGE CLASSIFICATION	Z0020/ADE(V)5-119 Z0020/ADE(V)5-119	1			

ISSUE NO: 4   REPAIR PARTS SCALE: 02060   PAGE NO: 12

* GRADE OF REPAIR (LIGHT, MEDIUM OR HEAVY)

GROUP GA 02060

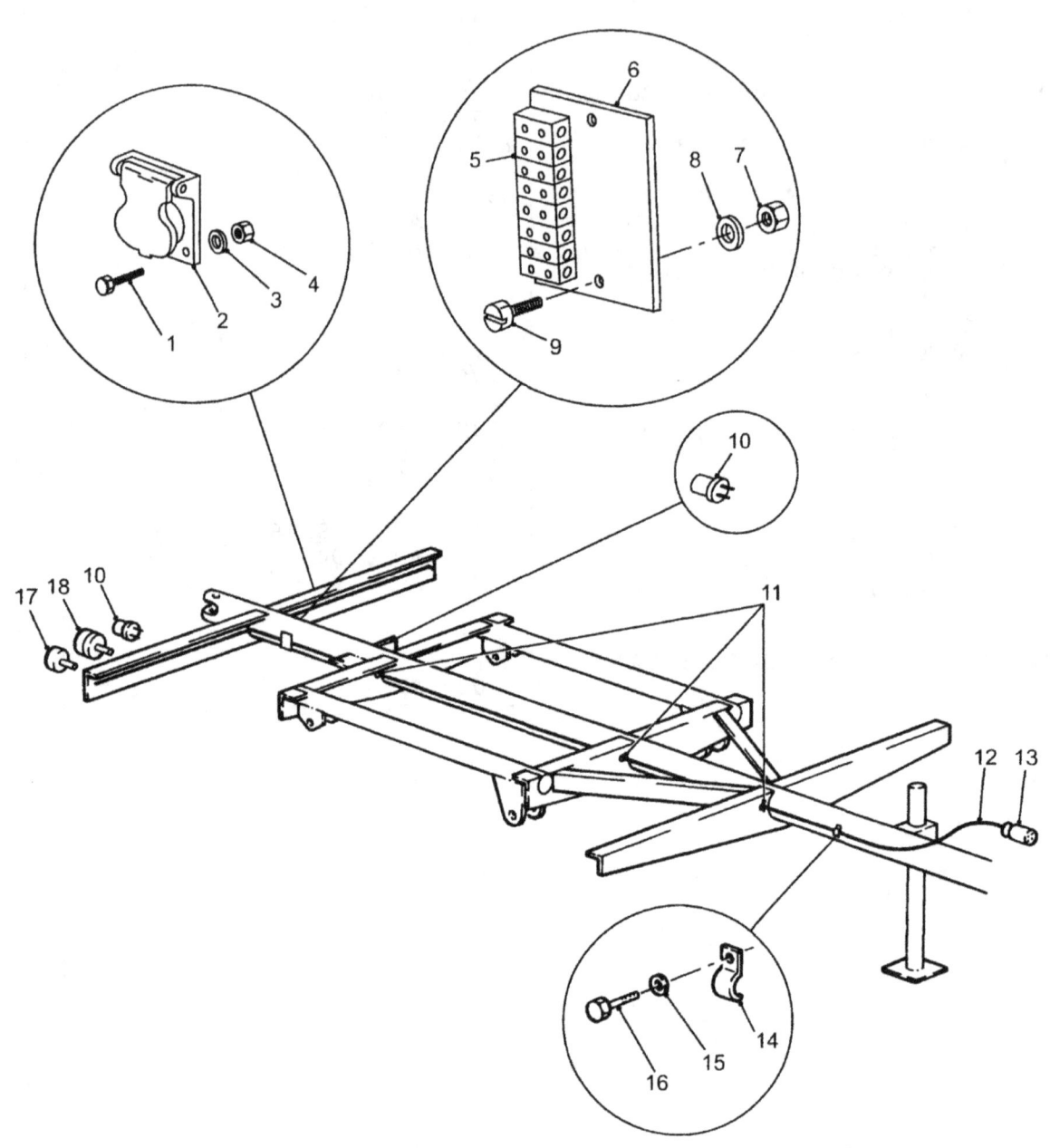

MAIN WIRING HARNESS

GROUP GA
TITLE MAIN WIRING HARNESS

ITEM NO	DESIGNATION	NSN MANUFACTURER CODE/PART NO SUPPLIER CODE/PART NO	NO OFF	U O I	E X P	L* M* H*
001	SCREW, CAP, HEXAGON HEAD	5305 01-411-3898 96906/MS90726-8 96906/MS90726-8	4 EA		X	LMH
002	CONNECTOR, RECEPTACLE, ELECTRICAL 12 CONTACTS, C/W SPRING LOADED COVER	5935 12-128-1833 Z0645/9203 Z0645/9203	1 EA		X	LMH
003	WASHER, LOCK STEEL, SPLIT HELICAL-RH, CAD PLATED, 9/32 IN. ID, 15/32 IN. OD, 11/16 IN. THK	5310 66-013-0067 Z4A36/20278 14MDSW Z4A36/20278 14MDSW	4 EA		X	LMH
004	NUT, PLAIN, HEXAGON UNF, 2B, STEEL, GRADE 0, ZINC COATED, 1/4 IN	5310 66-010-7421 Z0544/AS 2465 Z0544/AS 2465	4 EA		X	LMH
005	TERMINAL BOARD 12 TERMINALS, 152.40 MM LG BY 20 MM W BY 10 MM H, GREY	5940 66-091-9350 Z0732/LF12 Z1K57/BP3	1 EA		X	LMH
006	PLATE, INSTRUCTION TERMINAL STRIP, PLASTIC, YELLOW, 4-3/4 IN.LG BY 2-1/2 IN.W BY 0.145 IN.THK W/MTG HOLES	9905 66-025-2431 Z2813/ADE(V)5-67 Z2813/ADE(V)5-67	1 EA		X	LMH
007	NUT, PLAIN, SLOTTED, HEXAGON ETRIC, 6H, CLASS 8, STEEL, CAD PLATED AND PASSIVATED, 4 MM DIA, 7 MM W A/F, 5 MM H O/A	5310 66-098-4413 Z0544/AS1112 Z0544/AS1112	2 EA		X	LMH
008	WASHER, LOCK, 1/8 IN BOLT SIZE, SHAKEPROOF, EXTERNAL TOOTH, STEEL, ZINC PLATED, 0.18 IN THK (USE GA 906)	/NIL /NIL	2			
009	SCREW, MACHINE ISO METRIC, 6G, BRASS, FLAT FIL HD, SLOT DRIVE, 4 MM DIA, 25 MM LG	5305 66-098-4307 Z0544/AS 1427 Z0544/AS 1427	2 EA		X	LMH
010	LIGHT, PANEL MAP READING, CLEAR LENS, W/HOOD, W/O LAMP (USED ON EARLY AND LATE TYPE FRAMES, REFER GROUP GB)	Z2813/TSE(V)93-594 K1089/53189	1			
011	GROMMET, NONMETALLIC PLASTICS	5325 66-120-5419 Z0421/28R32 Z0421/28R32	3 EA		X	LMH

ISSUE NO: 4 REPAIR PARTS SCALE: 02060 PAGE NO: 13

* GRADE OF REPAIR (LIGHT, MEDIUM OR HEAVY)

GROUP GA 02060

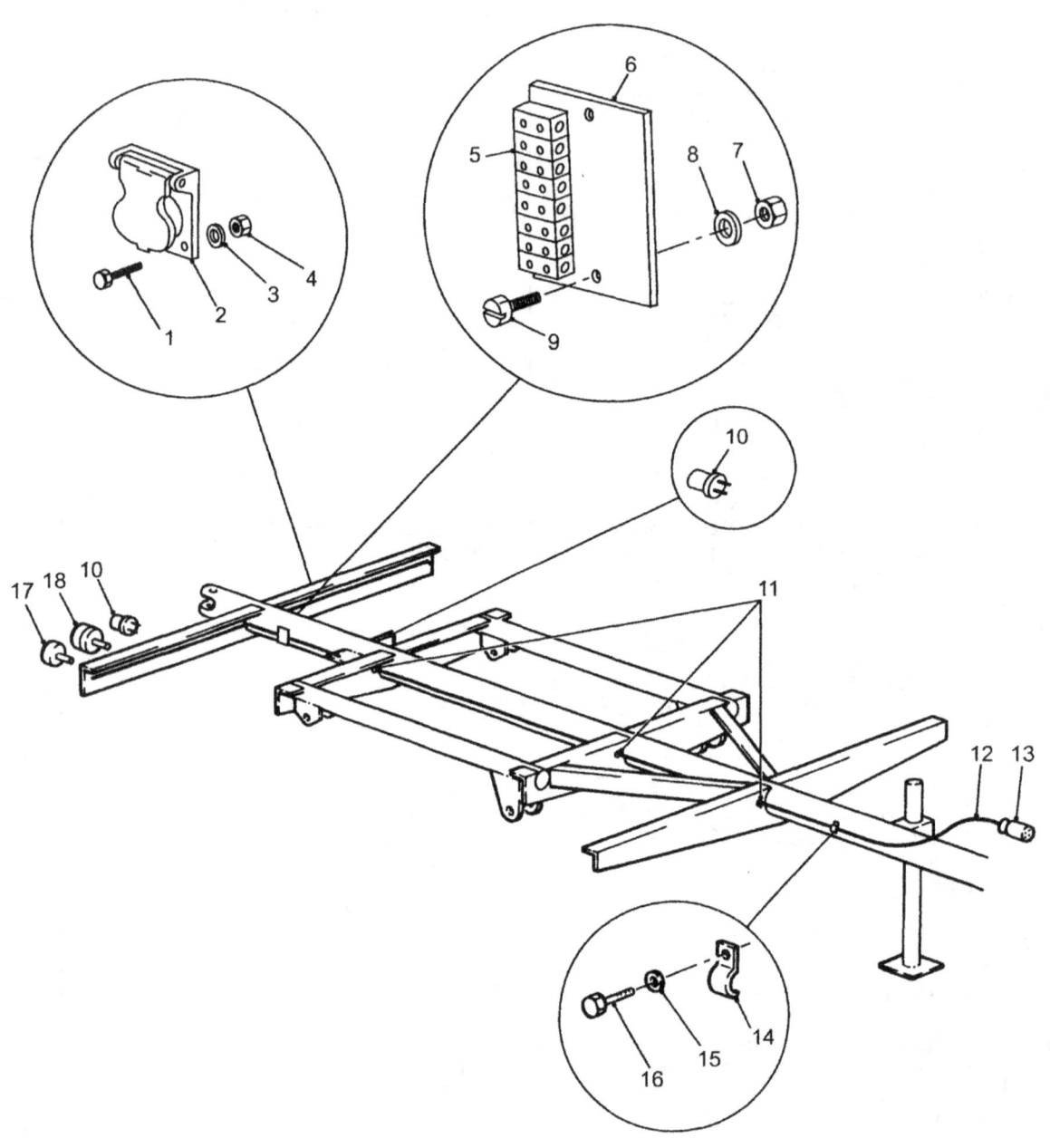

MAIN WIRING HARNESS

GROUP GA
TITLE MAIN WIRING HARNESS

ITEM NO	DESIGNATION	NSN MANUFACTURER CODE/PART NO SUPPLIER CODE/PART NO	NO OFF	U O I	E X P	L* M* H*
012	CABLE, POWER, ELECTRICAL 10 CONDUCTORS, 24/0.2 MM STRANDS, PVC INSULATION 8 WHITE, 1 BLACK, 1 GREEN/YELLOW, PVC SHEATH ORANGE 250 V , ROLL OF 100 METRES (USE AS REQUIRED)	6145 66-122-9502 Z0499/BFFR02AA009 Z0499/BFFR02AA009	12MR		X	MH
013	CONNECTOR, PLUG, ELECTRICAL MALE, 12 CONTACTS	5935 12-188-1750 Z0645/9204 Z0645/9204	1EA		X	LMH
014	CLIP, HALF SADDLE TY CLIP, ELECTRICAL, HALF SADDLE TYPE	5940 66-093-2784 Z0220/H896 Z0220/H896	1EA		X	LMH
015	WASHER, LOCK STEEL, CADMIUM PLATED, 3/16 IN.BOLT SIZE, SHAKEPROOF EXTERNAL OR NO10 SIZE)	5310 66-016-8516 Z1272/038-294A Z0088/1110	7EA		X	LMH
016	SCREW, TAPPING THREAD FORMING, STEEL, PAN HD, SLOTTED , DOG TYPE POINT, NO 6 BY 3/8IN	5305 66-016-0720 80204/B18-6-4 80204/B18-6-4	7EA		X	LMH
017	LIGHT, TURN, REAR, AMBER (REFER GROUP GB)	K1089/488 K1089/488	2			
018	STOP/TAIL LIGHT (REFER GROUP GB)	Z0645/2386 Z0645/2386	1			
901	STOP LIGHT-TAILLIGHT, VEHICULAR RED GLASS LENS, ROUND TAPERED SHAPE LIGHT, 12 V INCANDESCENT BULB (USED ON LATE TYPE FRAME)	Z0645/2351 Z0645/2351	2			
902	STRAP, RETAINING CONDUIT, HALF SADDLE, 20 MM, GALVANISED	5340 66-116-5535 Z0362/180/20 Z0220/H313-3/4	6EA		X	LMH
903	STRAP, RETAINING (QTY 2 OFF USED ON LATE TYPE FRAME)	5340 66-044-6204 Z0220/H1501 Z0220/H1501	13EA		X	LMH
904	STRAP, RETAINING (USED ON LATE TYPE FRAME)	5340 66-044-6203 Z0220/H2243 Z0220/H2243	11EA		X	LMH

ISSUE NO: 4   REPAIR PARTS SCALE: 02060   PAGE NO: 14

* GRADE OF REPAIR (LIGHT, MEDIUM OR HEAVY)

GROUP GA 02060

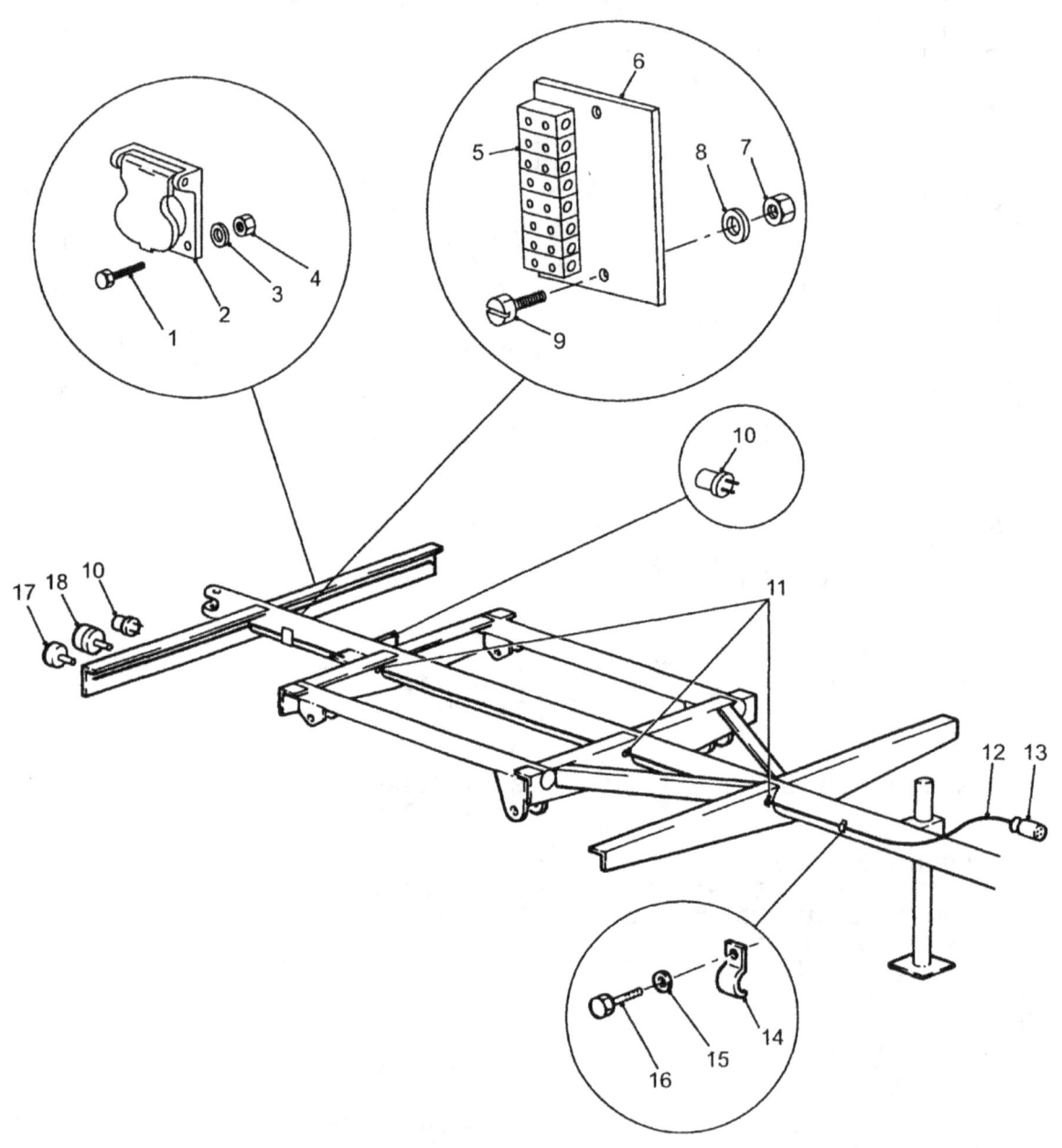

MAIN WIRING HARNESS

GROUP   GA
TITLE   MAIN WIRING HARNESS

ITEM NO	DESIGNATION	NSN MANUFACTURER CODE/PART NO SUPPLIER CODE/PART NO	NO OFF	UOI	EXP	L* M* H*
905	SPLICE SET, QUICK DISCONNECT RED, SINGLE CIRCUIT, 2 WIRE	5940 66-017-9989 Z0220/H850 Z5394/MYH1780	8	EA	X	LMH
906	WASHER, LOCK, 1/8 IN BOLT SIZE, SHAKEPROOF, EXTERNAL TOOTH, STEEL, ZINC PLATED, 0.18 IN THK, PACK OF 1000 (USE AS REQUIRED)	5310 66-147-0583 ZD484/02138501 ZD484/02138501	1	PK	X	LMH

* GRADE OF REPAIR (LIGHT, MEDIUM OR HEAVY)

GROUP GB SHEET 1 02060

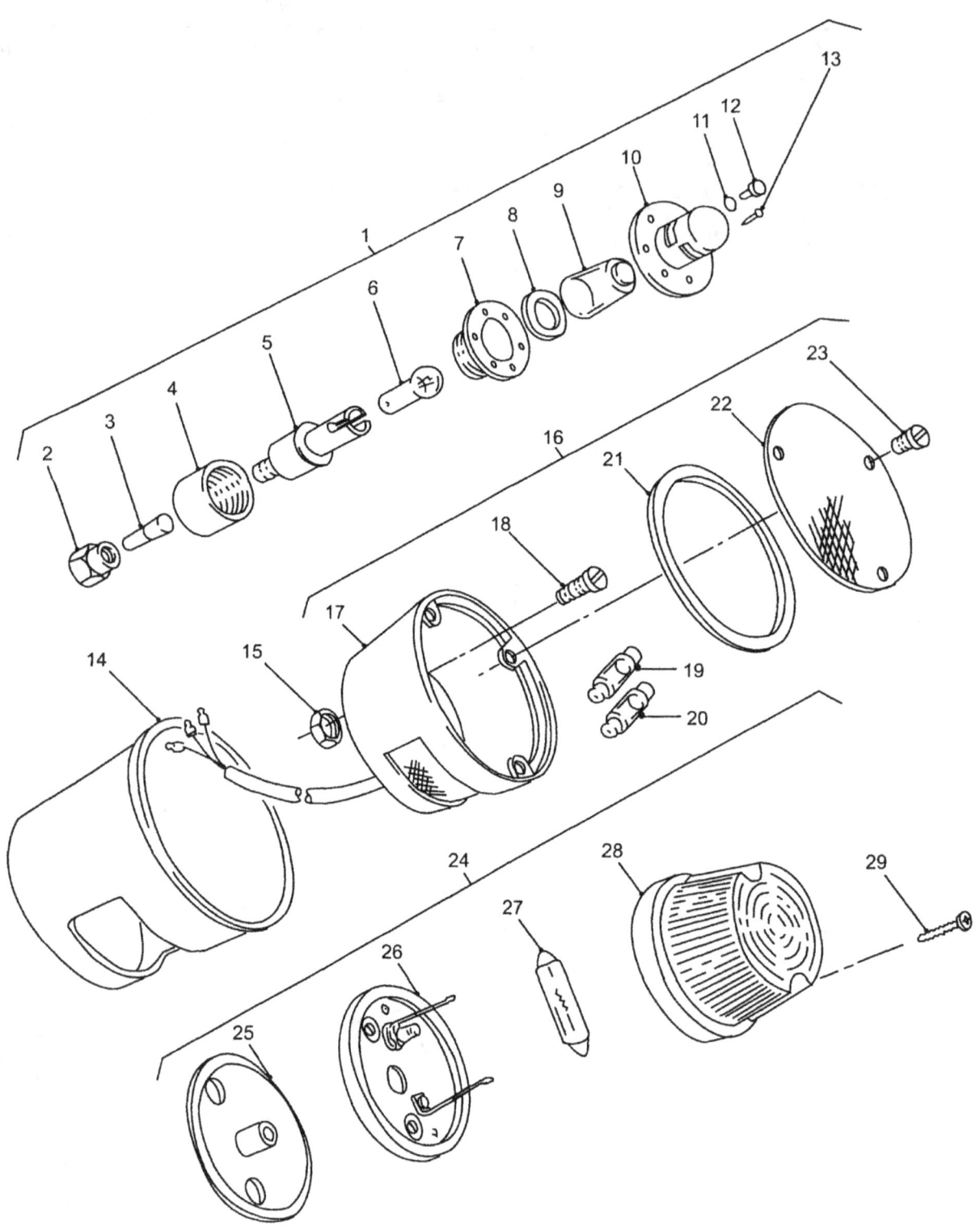

LIGHTS

GROUP GB/1
TITLE LIGHTS

ITEM NO	DESIGNATION	NSN MANUFACTURER CODE/PART NO SUPPLIER CODE/PART NO	NO OFF	U O I	E X P	L* M* H*
001	LIGHT, PANEL MAP READING, CLEAR LENS, W/HOOD, W/O LAMP (INCLUDES GB 002 TO GB 013, USED ON EARLY AND LATE TYPE FRAMES)	6210 66-011-0776 Z2813/TSE(V)93-594 ZA233/53189	1	EA	X	LMH
002	NUT (ALSO PART OF GB 001)	K0049/445721 K0049/445721	1			
003	FERRULE (ALSO PART OF GB 001)	K0049/573075 K0049/573075	1			
004	NUT, RING (ALSO PART OF GB 001)	K0049/573049 K0049/573049	1			
005	LAMPHOLDER, SUB ASSEMBLY (ALSO PART OF GB 001)	6220 99-803-3390 K1089/573085 ZA233/573085	1	EA	X	LMH
006	LAMP, INCANDESCENT 12 V, 5 W, SINGLE CONTACT BAYONET BA15S BASE, "G" SHAPE, CLEAR (ALSO PART OF GB 001)	6240 66-010-7460 Z0645/G125 Z0645/G125	1	EA	X	LMH
007	FLANGE (ALSO PART OF GB 001)	K0049/573040 K0049/573040	1			
008	WASHER, FLAT COVER TO BACK PLATE (ALSO PART OF GB 001)	5310 99-803-3383 K1089/573053 K1089/573053	1	EA	X	LMH
009	LENS, LIGHT GLASS, NUMBER PLATE LAMP (ALSO PART OF GB 001)	6220 99-803-3384 U1030/AAU7897 U1030/AAU7897	1	EA	X	LMH
010	SHIELD (ALSO PART OF GB 001)	Z5394/HYG5052 Z5394/HYG5052	1			
011	WASHER (ALSO PART OF GB 001)	K0049/188412 K0049/188412	3			
012	SCREW, MACHINE BRASS, HEX HD, NICKEL PLATED, NO 4 BY 5/16IN LG (ALSO PART OF GB 001)	5305 99-128-7539 U0795/FV396981-4 ZA212/36752	3	EA	X	LMH

ISSUE NO: 4   REPAIR PARTS SCALE: 02060

* GRADE OF REPAIR (LIGHT, MEDIUM OR HEAVY)

GROUP GB SHEET 1   02060

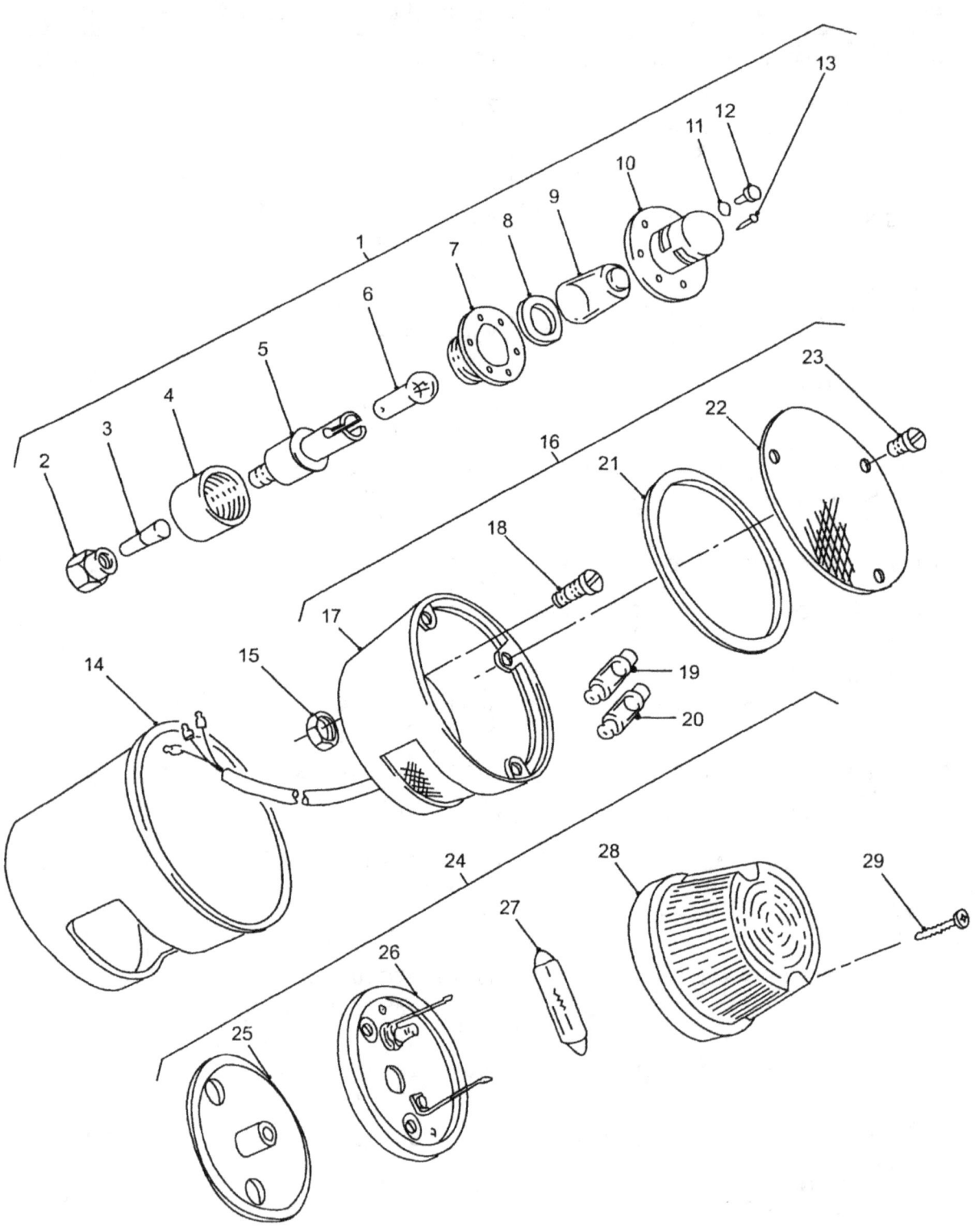

LIGHTS

GROUP GB/1
TITLE LIGHTS

ITEM NO	DESIGNATION	NSN MANUFACTURER CODE/PART NO SUPPLIER CODE/PART NO	NO OFF	U O I	E X P	L* M* H*
013	SCREW, MACHINE BA, MILD STEEL, CHEESE HEAD, SLOTTED, CAD PLATED, NO.4 BY 3/8IN (ALSO PART OF GB 001)	5305 66-012-5751 Z0544/B85 Z0544/B85	3	EA	X	LMH
014	COVER, TAILLIGHT	Z2813/ADE(V)225-752 Z2813/ADE(V)225-752	1			
015	SPACER	/NIL /NIL	2			
016	STOP LIGHT-TAILLIGHT, VEHICULAR (INCLUDES GB 017 TO GB 023, USED ON EARLY TYPE FRAME)	6220 66-090-7823 Z0645/2386 Z0645/2386	1	EA	A	LMH
017	LAMP HOLDER (USE GB 016)	/NIL /NIL	1			
018	SCREW, MACHINE UNF2A, BRASS, CHEESE HD, SLOT DRIVE, 1/4IN BY 1-1/4IN LG (ALSO PART OF GB 016)	5305 66-021-0731 ZA229/1/4X1-1/4UNFFIL/HEAD.BRASS.MTS ZA229/1/4X1-1/4UNFFIL/HEAD.BRASS.MTS	2	EA	X	LMH
901	NUT, PLAIN, HEXAGON UNF 2B, BRASS, 1/4IN (ALSO PART OF FAA 016)	5310 66-015-5038 Z0544/B147 Z0544/B147	2	EA	X	LMH
019	LAMP, INCANDESCENT 12V, 5 W, BRASS FERRULES BASE, "CYL" SHAPE, CLEAR (ALSO PART OF GB 016 AND GB 030)	6240 66-014-8509 Z0645/L125 Z0645/L125	1	EA	X	LMH
020	LAMP, INCANDESCENT 12 V, 18 W, BRASS FERRULE BASE, "CYL" SHAPE, CLEAR (ALSO PART OF GB 016 AND GB 030)	6240 66-027-7458 Z0645/K1218 Z0645/K1218	1	EA	X	LMH
021	GASKET TAIL AND STOP LIGHT LENS (ALSO PART OF GB 016)	Z0645/2.9806.05 Z0645/2.9805.05	1			
022	LENS, LIGHT RED, PLASTIC, 108 MM O/A DIA (ALSO PART OF GB 016)	6220 66-092-6259 Z0645/2.9801.01 Z0645/2.9801.01	1	EA	X	LMH
023	SCREW TAPPING (USE GB 016)	/NIL /NIL	3			

ISSUE NO: 4   REPAIR PARTS SCALE: 02060   PAGE NO: 17

* GRADE OF REPAIR (LIGHT, MEDIUM OR HEAVY)

GROUP GB SHEET 1   02060

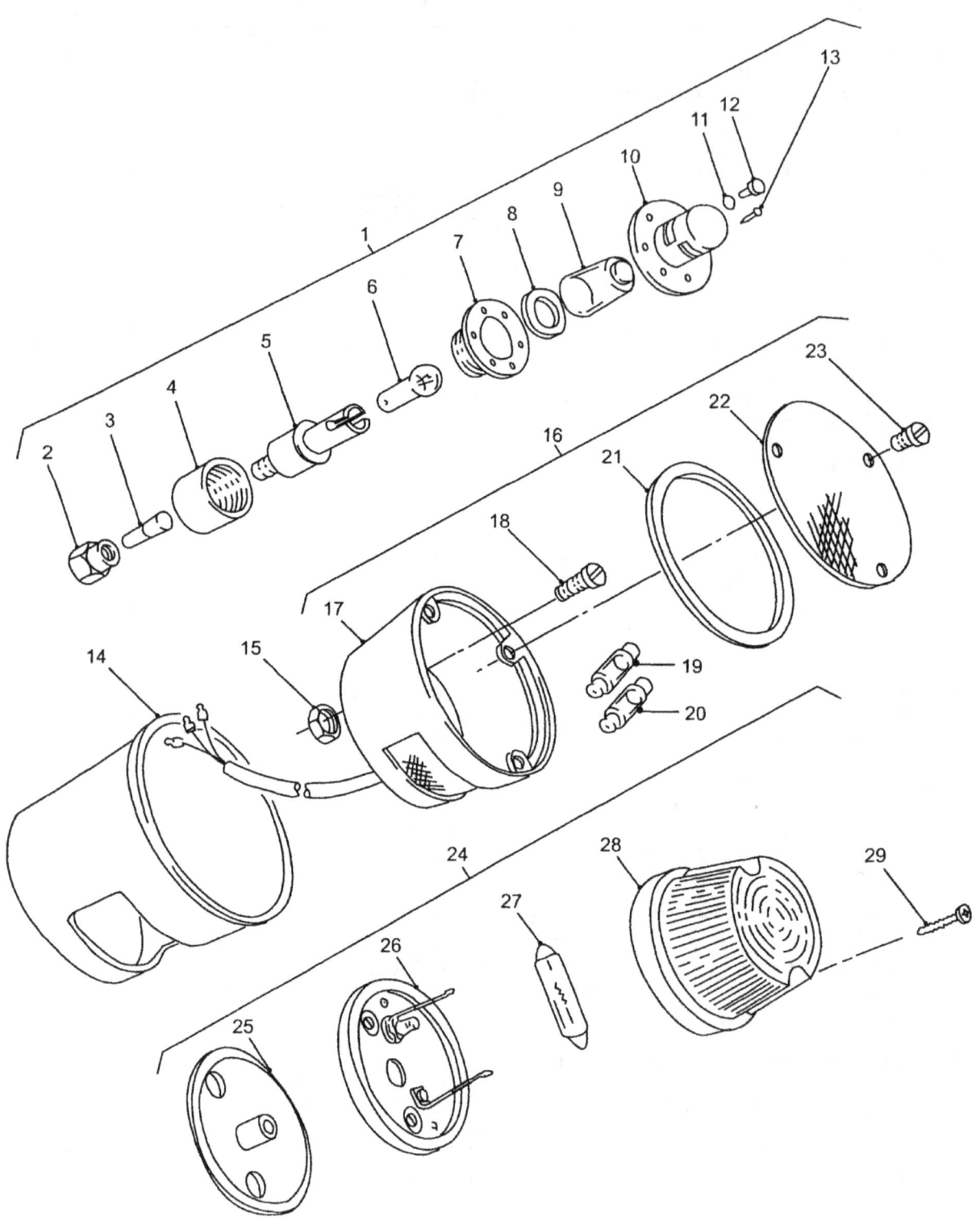

LIGHTS

GROUP GB/1
TITLE LIGHTS

ITEM NO	DESIGNATION	NSN MANUFACTURER CODE/PART NO SUPPLIER CODE/PART NO	NO OFF	UOI	EXP	L*M*H*
024	DIRECTIONAL LIGHT, VEHICULAR AMBER LENS, ROUND TAPERED SHAPE, 76 MM O/A DIA, 43.5 MM O/A LG, 12V OR 24V BULB (INCLUDES GB 025 TO GB 029, USE ON EARLY AND LATE TYPE FRAMES)	6220 66-143-7497 Z0645/2100 Z0645/2100	2 EA	X	LMH
025	BACKPLATE, RUBBER, CIRCULAR LAMP (ALSO PART OF GB 024 AND GB 030)	Z0645/2.0423.03 Z0645/2.0423.03	1			
026	GLOBE HOLDER (USE GB 024)	/NIL /NIL	1			
027	LAMP, INCANDESCENT 12V, 21W, 41 MM O/A LG, 15.5 MM O/A DIA (ALSO PART OF GB 024)	6240 66-143-7499 Z0645/K1221 Z0645/K1221	1 EA	X	LMH
028	LENS, LIGHT FRONT DIRECTION INDICATOR LAMP, AMBER, 76 MM DIA (ALSO PART OF GB 024)	6220 66-043-0011 Z0645/2.6830.01 Z0645/2.6830.01	1 EA	X	LMH
029	SCREW, SELF TAPPING, CROSSHEAD, 1/2 IN, 8 GAUGE (USE GB 024 OR GB 030)	/NIL /NIL	2			

ISSUE NO: 4   REPAIR PARTS SCALE: 02060   PAGE NO: 18

* GRADE OF REPAIR (LIGHT, MEDIUM OR HEAVY)

GROUP GB SHEET 2　　　　02060

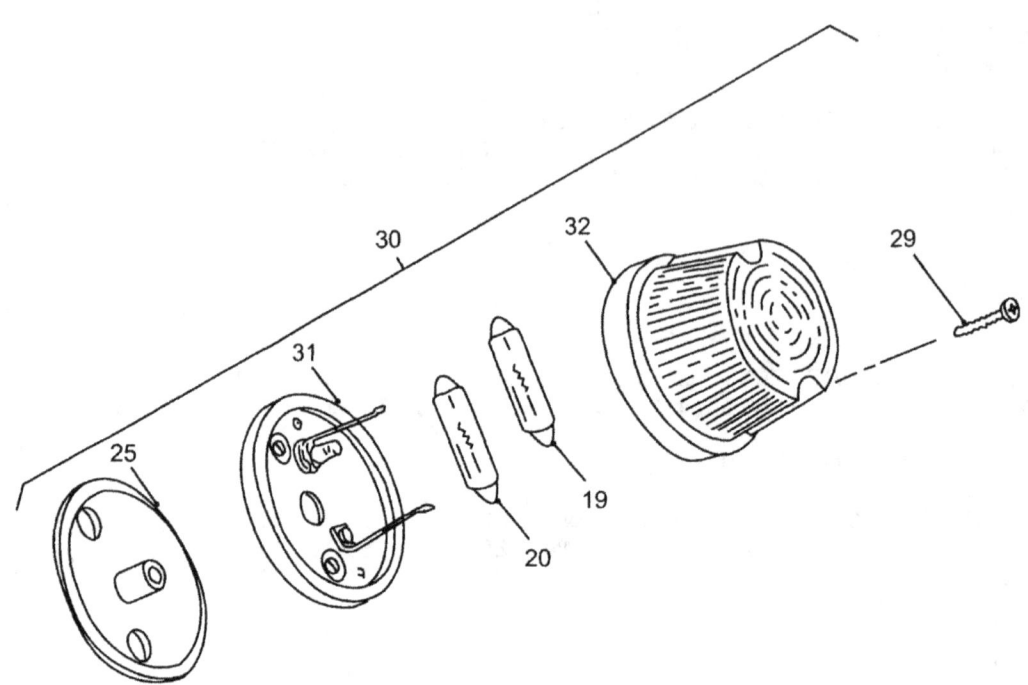

LIGHTS

GROUP GB/2
TITLE LIGHTS

ITEM NO	DESIGNATION	NSN MANUFACTURER CODE/PART NO SUPPLIER CODE/PART NO	NO OFF	U O I	E X P	L* M* H*
030	STOP LIGHT-TAILLIGHT, VEHICULAR RED LENS, ROUND TAPERED SHAPE, 76 MM O/A DIA, 43.5 MM O/A LG (INCLUDES GB 019, GB 020, GB 025, GB 029, GB 031 AND GB 032, USED ON LATE TYPE FRAME)	6220 66-143-7496 Z0645/2351 Z0645/2351	2 EA		X	LMH
031	GLOBE HOLDER (USE GB 030)	/NIL /NIL	1			
032	LENS, LIGHT RED, ROUND DOME SHAPE, 43.5 MM O/A LG, 76 MM O/A DIA, FLANGE MTG (ALSO PART OF GB 030)	5850 66-143-7498 Z0645/2.2885.01 Z0645/2.2885.01	1 EA		X	LMH

ISSUE NO: 4   REPAIR PARTS SCALE: 02060   PAGE NO: 19

* GRADE OF REPAIR (LIGHT, MEDIUM OR HEAVY)

INTENTIONALLY LEFT BLANK

www.ingramcontent.com/pod-product-compliance
Lightning Source LLC
Chambersburg PA
CBHW080028130526
44591CB00037B/2707